T0301384

Socioecological Transitions and Global Change

ADVANCES IN ECOLOGICAL ECONOMICS

Series Editor: Jeroen C.J.M. van den Bergh, *Professor of Environmental Economics, Free University, Amsterdam, The Netherlands*

Founding Editor: Robert Costanza, *Director, University of Maryland Institute for Ecological Economics and Professor, Center for Environmental and Estuarine Studies and Zoology Department, USA*

This important series makes a significant contribution to the development of the principles and practices of ecological economics, a field that has expanded dramatically in recent years. The series provides an invaluable forum for the publication of high-quality work and shows how ecological economic analysis can make a contribution to understanding and resolving important problems.

The main emphasis of the series is on the development and application of new original ideas in ecological economics. International in its approach, it includes some of the best theoretical and empirical work in the field with contributions to fundamental principles, rigorous evaluations of existing concepts, historical surveys and future visions. It seeks to address some of the most important theoretical questions and gives policy solutions for the ecological problems confronting the global village as we move into the 21st century.

Titles in the series include:

The Economics of Technology Diffusion and Energy Efficiency
Peter Mulder

Time Strategies, Innovation and Environmental Policy
Edited by Christian Sartorius and Stefan Zundel

America's Changing Coasts
Private Rights and Public Trust
Edited by Diana M. Whitelaw and Gerald R. Visgilio

Economic Growth, Material Flows and the Environment
New Applications of Structural Decomposition Analysis and Physical
Input–Output Tables
Rutger Hoekstra

Joint Production and Responsibility in Ecological Economics
On the Foundations of Environmental Policy
Stefan Baumgärtner, Malte Faber and Johannes Schiller

Frontiers in Ecological Economic Theory and Application
Edited by Jon D. Erickson and John M. Gowdy

Socioecological Transitions and Global Change
Trajectories of Social Metabolism and Land Use
Edited by Marina Fischer-Kowalski and Helmut Haberl

Socioecological Transitions and Global Change

Trajectories of Social Metabolism and Land Use

Edited by

Marina Fischer-Kowalski and Helmut Haberl

Institute of Social Ecology, Klagenfurt University, Vienna, Austria

ADVANCES IN ECOLOGICAL ECONOMICS

Edward Elgar
Cheltenham, UK • Northampton, MA, USA

Published by
Edward Elgar Publishing Limited
Glensanda House
Montpellier Parade
Cheltenham
Glos GL50 1UA
UK

Edward Elgar Publishing, Inc.
William Pratt House
9 Dewey Court
Northampton
Massachusetts 01060
USA

A catalogue record for this book
is available from the British Library

Library of Congress Cataloging in Publication Data

Socioecological transitions and global change : trajectories of social
metabolism and land use / edited by Marina Fischer-Kowalski and Helmut
Haberl.
    p. cm. – (Advances in ecological economics series)
    Includes bibliographical references and index.
    ISBN-13: 978-1-84720-340-3 (hardcover : alk. paper)
    1. Social ecology. 2. Social evolution. 3. Social change. 4. System theory.
    5. Globalization–Economic aspects. 6. Globalization–Environmental
    aspects. I. Fischer-Kowalski, Marina, 1946- II. Haberl, Helmut.
    HM861.S63 2007
    304.2–dc22
                                                                    2007000703

ISBN 978 1 84720 340 3

Printed and bound in Great Britain by MPG Books Ltd, Bodmin, Cornwall

Printed with support of the Research Council of Klagenfurt University and the research programme 'Kulturlandschaftsforschung' of the Austrian Ministry of Education, Science and Culture. We thank Ursula Lindenberg for improving the language and her help with the formatting of the manuscript.

We acknowledge continuous support from the Faculty for Interdisciplinary Studies, Vienna, in particular its Dean, Roland Fischer, and its Librarian, Bernhard Hammer. We also wish to mention some people who have been a valuable source of inspiration and only by chance did not become co-authors of this book: Veronika Gaube, Thomas Macho, Rolf Peter Sieferle, Barbara Smetschka, Helga Weisz and Verena Winiwarter.

Contents

Figures

Tables

Contributors

Nina Eisenmenger, Institute of Social Ecology, Klagenfurt University, Vienna, Austria

Karl-Heinz Erb, Institute of Social Ecology, Klagenfurt University, Vienna, Austria

Marina Fischer-Kowalski, Institute of Social Ecology, Klagenfurt University, Vienna, Austria

Clemens M. Grünbühel, the Mekong Institute, Khon Kaen, Thailand

Helmut Haberl, Institute of Social Ecology, Klagenfurt University, Vienna, Austria

Fridolin Krausmann, Institute of Social Ecology, Klagenfurt University, Vienna, Austria

Jesus Ramos Martin, Institute for Environmental Science and Technology, Autonomous University of Barcelona, Barcelona, Spain

Heinz Schandl, Commonwealth Scientific and Industrial Research Organisation, CSIRO, Sustainable Ecosystems, Resource Futures Programme. Canberra, Australia

Simron J. Singh, Institute of Social Ecology, Klagenfurt University, Vienna, Austria

Foreword

This is a book of human ecology, environmental history, and ecological economics that emerged over the last fifteen years in a pioneering and successful research programme led by Marina Fischer-Kowalski in Austria. One cannot but admire, as I do, the persistence of the leader of this group in balancing the social sciences and the natural sciences while developing this line of research, and rejoice in her success in attracting a brilliant team of younger researchers such as Helmut Haberl, Fridolin Krausmann, Karl-Heinz Erb, Heinz Schandl, Helga Weisz, and others now coming from other countries. The results are obvious in the many publications in journals, such as *Ecological Economics* and *The Journal of Industrial Ecology*, in addition to books in both German and in English. This volume is perhaps the best one, a collection of chapters on Europe, Asia and Latin America analysing transitions in the use of energy and materials, patterns of human time-use and economic changes, with the intention of explaining the past, but also aiming at sorting out possible and impossible futures.

The economy may be seen as a system of transformation of energy and materials into products and services for human consumption, and ultimately into dissipated heat, carbon dioxide, and solid or liquid wastes. The study of the 'metabolism of society' requires specific methodologies of accounting that have become standardized only in the last ten years or so. Thus, Eurostat now regularly publishes statistics on the materials flows (in tons per year) of the countries of the European Union. Such statistics allow us to trace historical patterns, make comparisons, and discuss and dismiss the hypothesis of 'dematerialization' of the economy.

Different patterns of socioeconomic metabolism also imply different patterns of land use. For instance, if fossil fuels are substituted for fuelwood, forests will perhaps be used less intensively, thus having a chance to recover from past overuse. By contrast, the addition of substantial amounts of biodiesel or bioethanol to the energy system will bring an increase in the human appropriation of biomass, probably to the detriment of other species.

Different countries show different trajectories in their trade balances of energy and materials. For instance, as analysed in some chapters of this splendid book, the United Kingdom exported energy before 1914 in the form of coal. Again, in the 1980s and 1990s the United Kingdom became at least self-sufficient in oil and even became an exporter before reaching the North-Sea oil extraction peak a short time ago. Meanwhile, Austria followed a historical pattern common to many other European continental countries, becoming a larger and larger net importer of energy especially from the 1950s onwards.

There are differences in the histories of the metabolism of societies, but the main point of this book is that there is much that they have in common. Industrialization and economic growth bring countries to levels of energy use per year and per capita of 200 GJ or more, with 12 tons or more of material flows per year and per capita. China and India are still behind on this road towards higher per capita levels of use of energy and materials, but they are travelling the same road. There are certainly variations between such averages, however, there is an enormous jump between agrarian poor countries (based on biomass), and industrialized rich countries, as shown in the research in this book on some small scale rural societies in South East Asia by the ecological anthropologist Clemens Grünbühel (a specialist on Laos), and by the economic and environmental historian Simron J. Singh (a specialist on the Nicobar islands).

Does economic growth lead (with today's technologies and consumption patterns) to a rather similar common future of the world societies as regards time-use and the metabolic flows? The answer is, alarmingly, affirmative. The demographic transition will be completed at a level of perhaps nine billion people in the world. But the 'socio-economic metabolic transition' will continue, in terms of use of energy and materials (including water) and a decreasing share of active agricultural population, towards a stage where humankind (by using natural gas, coal, and the remaining oil), will produce greenhouse gas emissions three or even four times larger than today.

A foreseeable but an impossible future.

Joan Martinez-Alier
Department of Economics and Economic History,
Universitat Autonoma de Barcelona and
President of the International Society for Ecological Economics
(2006, 2007)

1. Conceptualizing, observing and comparing socioecological transitions

Marina Fischer-Kowalski and Helmut Haberl

1.1 INTRODUCTION

A transition to a more sustainable state of society and the environment, a perspective that envisages attractive human futures on a hospitable planet Earth – this is a vision that nowadays inspires much research and policy-making. The notion of transition implies a major change – not incremental adjustments or improvements, but a qualitatively new state of the system. Transitions of a different kind may well be under way already, however. Do we not experience a rapid, even increasing pace of change in our working lives, our families, many of our institutions, our technologies and our everyday culture? Do we not perceive rapid transformations of landscapes in industrial centres as well as in holiday resorts at the periphery? And do we not have the impression that the weather has changed compared with our childhood, invalidating old rules of thumb? On the timescale of human lives, which is the common timescale for comparing experiences, time does not stand still at all. It seems rather the case that people are finding it difficult to move as fast as the world around them.

The environmental historian, John McNeill (2000), addressed this phenomenon in the ironically titled publication, *Something New Under the Sun*, a review of the 20th century. According to the statistics he assembled, there is barely any dimension of human social life and interference with the environment that has not undergone a rapid expansion worldwide during this one century, an expansion that has exceeded the factor 5 growth of the human population, substantial in itself, sometimes by an order of magnitude.[1] In this context, what makes us believe that we exist in some steady state or equilibrium? How could we possibly not think that we are in the middle of a rather explosive transition?

Usually, 'transitions' are broken down into a formal sequence of phases that lead from a status quo state to a new equilibrium. A common distinction

1

is made between a *take-off phase* (in which the status quo is still in place, but various symptoms of its initial destabilization are evident), followed by an *acceleration phase* in which many rapid changes take place and a subsequent *stabilization phase* in which those changes are slowing down and the features of a new equilibrium are beginning to crystallize. If the particles involved in such a transition had some consciousness, how would they feel during each phase? If we mobilize our empathetic imagination, we might conjure the following pictures. In the take-off phase, the particles would experience a destruction of order. To some of them, this would appear very threatening. To others, however, this might appeal as a form of liberation from traditional constraints. Some particles therefore would feel the need to fight in defence of that order, and some would organize to accelerate its destruction – controversies and struggles would arise. What then should we expect in the acceleration phase of a transition? The most probable mental state of the particles would be one of passivity: the feeling of being driven in some direction without having a chance to do much about it, except for remaining mobile so as not to be trampled in the process. Those who are further in front would appear to be faster, as leaders to be followed, and anyone slowing down would be kicked aside. In the stabilization phase, finally, one would expect there to be great insecurity as to what the new state was going to be like. Again, struggles might be expected to ensue over the features of the new order and the interpretation of experiences various particles have made.

If we take John McNeill's reconstruction of the 20th century seriously, we are neither in a stable status quo, nor in the state of a new equilibrium. If some of the conclusions from the classical Club of Rome study on the limits to growth (Meadows, Meadows and Randers 1972) may have been wrong, one presumption was certainly correct: we have just one planet at our disposal (so far at least). An ongoing explosion within the confines of a limited space cannot be described as an equilibrium, even, if we use the term 'dynamic' very liberally. At the same time, it is also hard to believe that we might still be in a take-off phase, since the entire past century was marked by such tremendous changes. It appears most likely that we are within an acceleration phase, in the middle of a transition. How far we still have to go in this acceleration remains unclear. Possibly, we are already close to a new equilibrium, since it is highly improbable that the socioecological regime we see now can continue for, let us say, another 200 years. The ongoing transition is bound to lead to some new, as yet unknown state. How pleasant this new state will be for humans remains to be seen and it is not known how quickly such a new equilibrium will be reached after a potential period of massive disturbances.

One may develop such a scenario regarding transitions from the vantage point of environmental history. The particular historian we refer to

(McNeill 2000) focuses on a large number of variables and on a time span covering centuries. Given another timescale, and using another selection of variables, other events may appear as transitions. The choice of timescale to be employed depends among other things on preconceptions concerning sustainability: if one feels a sustainable state of society and the environment is fairly close to where we are now, there is no need to discuss time spans of more than a few decades. If, however, one feels that such a state is significantly different from the present and will therefore take quite some time to develop – as the authors of this publication tend to believe – then one needs to look at larger time spans. The specific choice that moulds this book's approach involved analysis of a time span that looks back to the last major transition in world history: the transition from agrarian to industrial society. At various times, at various places and at various scales, we may observe different phases of this transition process, a process that is still ongoing.

1.2   A FAMILY OF TRANSITION CONCEPTS

A specific feature of transition is the idea that it lies between two qualitatively distinct states, and that no linear, incremental path leads from one state to the other but rather a dynamic, possibly chaotic process of change. Looked at from the inverse perspective, it is the particular methodological (even epistemological) achievement of the transition notion that it allows two qualitatively different states to be distinguished (naming them differently, for example). In comparison, theories of growth or modernization do not provide any order except for that defined by the time axis: one may refer to an 'early' or 'later' period for example, while still assuming a certain homogeneity of the basic setting, and gradual change over time.[2] The 'transition' assumption of discontinuities along the time axis, by contrast, suggests a change in the basic setting.

   One has to be aware, however, that this perspective is extremely sensitive to the choice of scale. Consider the process of walking: from a wider perspective (in terms of time, say minutes, in terms of space, say landscape level), this appears as a continuous process, progressing in linear fashion across space, sometimes a little slower, sometimes faster. From a closer perspective (measured in seconds, with a spatial focus on the human body), it appears to be a cyclical process in which each leg is alternately lifted and set down on the ground. From an even closer perspective (measured in fractions of seconds, and observing groups of muscles) it would appear to be a transition process between one group of muscles contracting and then relaxing, and then another group of muscles in action. This demonstrates that transition theories and theories based on the assumption of gradualism do not necessarily contradict each other. One type of process may well be 'nested' within the other. These only compete when both types of theories are expected to explain a given phenomenon on the same spatial and temporal scale.[3]

Another ingredient of the transition notion is the idea of spontaneity or emergence: it is neither possible for one state to be deliberately transformed into the other, nor for the process to be fully controlled. Particularly if, as in sustainability research, the concept is applied to complex systems (such as societies or technology regimes), one is dealing with autocatalytic or autopoietic processes (Varela, Maturana and Uribe 1974) to which concepts of orderly governance, steering or management cannot be applied. It is commonly assumed, however, that the system's resilience is decreased during the take-off phase of a transition, when the old interrelations are breaking apart but no clear directionality of change has yet been established (Berkhout, Smith and Stirling 2003; Rotmans, Kemp and van Asselt 2001). This lack of resilience can present opportunities for competing forces within the system (as is assumed, for example, in Marxist theories), and it provides improved opportunities for systemic intervention. Systemic intervention takes into account the self-regulatory quality of complex systems, seeks to disturb them and cause resonance leading to change from within (for social systems, see Willke 1996).

Systems theorists have observed the phenomenon that complex systems, in particular ecosystems, may have a number of stable states or equilibria, that is, states of being to which the system will return after periods of disturbance (May 1977). When a system is moved beyond a threshold ('tipping point') it may rapidly flip from one equilibrium to another qualitatively different one – a process often described as 'regime shift', although other notions such as transition or transformation are also sometimes used. Systems are defined as resilient when they respond smoothly to gradual changes and return to their original state after disturbance. When approaching a threshold, however, a system is considered to be unstable, implying that it is undergoing rapid and largely unpredictable change. Small changes in conditions may then result in qualitatively different trajectories.

This concept, originally developed in general systems theory and often applied to ecosystems (for example, Gunderson and Holling 2002; Holling 1973; Scheffer et al. 2001), has recently been extended to socioecological systems, mainly by scholars assembled in the so-called Resilience Alliance (http://www.resalliance.org). Their basic concept is cyclical and may be summarized in the so-called 'lazy eight' diagram (for example, Gunderson and Holling 2002, p. 34). According to their view, change in ecosystems (and socioecological systems) follows a four-stage scheme. A stable, highly connected, mature system state that may last over long periods is thought to be followed by three transient phases in orderly sequence that taken together may be thought of as a transition. First, long-accumulated capital (biomass in the case of an ecosystem) is 'released' in an act of 'creative destruction' (a term borrowed from Schumpeter 1950). Second, the system is reorganized, for example, through innovation or restructuring. In a third phase, the system is dominated by quickly moving 'pioneers' that are then gradually replaced by large, slow, highly competitive units that succeed in monopolizing resources and again build up large capital stocks of a new, highly connected and stable state. In the remainder of this book we choose not to distinguish these three transient phases largely derived from observations of changes in ecosystems, since for our purposes, the three-phase scheme outlined above (take-off – acceleration – stabilization) derived from a social-scientific perspective was felt to be more straightforward.[4]

In any case, this discussion shows that the way in which the overall process, and the relations between the stages distinguished and connected by transitions, are conceptualized may make a big difference. Processes in complex living systems are usually assumed to be irreversible, so there is a clear directionality of time. This directionality can either imply consecutive stages of a 'developmental' type (like ontogenesis, or Herbert Spencer's notion of evolution, or Marxist historical materialism), or it may instead comply with a Darwinian type of evolution, assuming the future to be contingent upon the past, yet principally unpredictable.[5]

In the first case, when a developmental conceptual model is employed, each stage necessarily follows from the previous one and is, as a rule, considered superior or more 'mature' (as an adult is more mature than a child). Progression to this more mature stage can only be accelerated or delayed and if the process of maturing is severely impaired or prevented altogether, some unhealthy, handicapped state or even decay of the entire system will ensue. Thus, there is no real choice as to where to go from state I since whatever action is taken, state II will follow sooner or later, unless the system dies off altogether.

In the contemporary family of transition concepts, quite a few follow this developmental, Spencerian conceptual model. This applies, for example, when the World Bank and OECD refer to 'economies in transition': this phrase denotes economies in transition from (presumably outdated) state-planned to liberal market economies.[6] Similarly, it is assumed in developmental economics that countries and world regions still relying on agrarian subsistence have no choice but to undergo a transition to industrialized market economies. This is substantially in tune with the 19th-century belief in socioeconomic progress as embodied in Marx and Engels' writings about the inevitability and superiority of capitalism over feudalism, even if it falls short of their subsequent argument about the inevitability of socialism. In such developmental socioeconomic theories, similar to child development, transitions occur according to an endogenous process and the built-in dynamics of the system itself. According to this approach, the social and also possibly the natural environment do play a role in providing more or less favourable conditions, but they are not the drivers of the ongoing dynamic process. Therefore, deliberative interventions into such a process only play a small role: they may remove internal or external obstacles to growth and maturation, or they may build up obstacles. There is no theoretical guidance for choosing or influencing the direction of the process: this direction is already deeply embedded in the process itself.

In the second case, when either a more Darwinian evolutionary model or unspecified, open-ended change processes are assumed, the task of intervening into transition processes becomes more complex as it also encompasses the choice of direction. Given the objective difficulties of analysis, there is always the temptation to suggest that the proposed direction is inevitable (as in the Spencerian mind model above). Historians who have analysed structural change in the past may find the subjective conviction of actors makes little difference: even humble conceptions of human governance over major historical change (which appears to be simultaneously over-determined and under-determined) seem in retrospect to be overstated. But for the contemporary citizen who wishes to orient his or her personal, political and economic actions in an appropriate way, the interpretation of ongoing change and its direction does make a difference.

Contemporaries of the 'great transformation' (Polanyi 1944) from an agrarian to an industrial society, in the 18th and 19th centuries, clearly tended to share the perspective on evolutionary progress and its necessity that we have attributed to Herbert Spencer above. This applies to Adam Smith, who in his *Wealth of Nations* (1776) interpreted all of human history teleologically as a mere prelude to the emergence of a capitalist liberal market economy (or the mode of subsistence 'commerce', as he termed the stage following 'agriculture'), seen as the final form of civilization realizing human nature. For Adam Smith, there was no doubt that 'commerce' was superior to 'agriculture', and that it would triumph by virtue of its own dynamics. Deliberative human intervention could only have the effect of either producing obstacles slowing down this process or of removing such obstacles.

Karl Marx also shares Smith's enthusiasm for the superiority of capitalism over the previous stage (as he termed it, 'feudalism'), as well as his conviction that capitalism would finally sweep away the remnants of traditional society. For Marx, though, human agency plays a much larger role. This applies to the transition from feudal to capitalist society: even if the driving force in the background is technology (the 'development of the means of production'), it requires the bourgeoisie to be successful in class struggle for an actual breakthrough of the new social formation to be achieved. Equally, it requires not only the existence, but also the political organization of a new class acting towards the goal of overcoming capitalism. The next stage, socialism, would then be characterized not by the reign of blind market forces, but by collective human rationality and justice. This transition would not ensue mechanically,[7] nor could it be managed and controlled deliberately, but it would eventually come about as a result of societal struggles.

Nowadays, the discourse concerning a possible transition to a more sustainable form of society is no longer guided by trust in the iron necessity of progress. While a more sustainable society is generally considered superior to the current state, no strong impulse of history is felt that would take us there, whether we like it or not (Brand 1997). Nor are there highly organized conflicting societal forces that would drive this transition process. The transition to a more sustainable society is more a matter of reason than of passions, and certainly does not yet appear to be the logical and inevitable next stage. To many, though, it appears as the only plausible alternative to chaotic and possibly catastrophic developments of history. A sustainability transition is not conceived of as happening automatically, all by itself. It may only be brought about by deliberative human agency and this human agency may be organized in a variety of ways.

One approach towards understanding transitions, which focuses on the possibility of deliberative change towards sustainability, has been developed by Rip and Kemp (1998) and then elaborated by Rotmans et al. (2001) and Martens and Rotmans (2002). Their approach seems to have been fairly successful in informing and guiding the economic and technology policies of the Dutch government (Kemp and Loorbach 2006). The fundamental idea behind this involves exploring the possibility of technological change in certain niches, supporting seemingly more sustainable niche technologies by government policies and thereby hoping to stimulate a gradual change in the respective technology regime towards a new, more sustainable level. Berkhout et al. (2003) have elaborated on this approach by searching for 'quasi-evolutionary' ways in which technology regimes could be changed, not just through resonance on innovations in niches, but through policy interventions either via the 'selection pressures' on the regimes themselves (via energy taxation, for example) or via the 'adaptive capacity' of the network of actors collaborating within a technology regime.[8]

At present, this 'technological transitions' approach seems to be a politically operational strategy aiming to force technological systems towards sustainability. It is obviously more ambitious than mere attempts at cleaner production, even if those go as far as implying integrated process changes. This approach therefore comes closer to what we termed a socioecological transition (see below for more detail), but it still only deals with a fairly narrow range of variables. It distinguishes between three levels (micro – meso – macro). The micro level refers to technologies. The meso level deals with 'technology regimes' (which may in fact be several layers of nested regimes; see Berkhout et al. 2003) involving infrastructures, interconnected actors and financing systems. However, these notions of 'regimes' and 'regime change' signify societal features and changes less fundamental than our use of the term 'socioecological regime'. On the macro level, the approach talks about 'landscapes' to refer to a wider societal setting. While the original version focused mainly on a bottom-up transformation process (starting from the micro level with successful innovations in technological niches to then succeed in a gradual regime change, with resonance on the macro 'landscape' level), later criticism and extensions led to the specification of a whole variety of transformation processes, bottom-up or top-down, deliberate or spontaneous (Geels 2004; 2006). In all cases, though, this approach is restricted to time frames of no more than a few decades.

This book is based on a still more widely encompassing concept of transition – one that is focused on even larger time frames, from decades to centuries. We cannot easily deal with actors and their deliberate efforts on this temporal scale. The analysis focuses largely on structural change of interlinked social and natural systems. In theory at least, this refers to a much broader range of variables. In practice, though, we limit ourselves to a relatively narrow set, circumscribed by a particular paradigm as specified below, a set of variables localized at the society–nature interface for which quantitative measurements can be reliably obtained in very different contexts. The advantage of this self-restraint is that we can demonstrate the interconnectedness of (some) socioeconomic changes and (some) changes in natural systems very clearly, and thereby acquire the ability to model important necessities and constraints related to a sustainability transition.

1.3   CHARACTERIZING SOCIOECOLOGICAL TRANSITIONS

A socioecological transition, as we understand it, is a transition from one socioecological regime to another. How may we then define a socioecological regime? A socioecological regime is a specific fundamental pattern of interaction between (human) society and natural systems. To endow this sentence with meaning, we have to make at least a brief attempt to clarify the term 'human society' and to explain how it relates to natural systems.

(Human) Society and How it Relates to Natural Systems

How may we define a 'human society'? No clear consensus about this term exists, either within social science disciplines or between them. For the remainder of our argument, we shall need to relate human society (here-after referred to as 'society', for brevity's sake) to its natural environment, or, in more fundamental terms, to nature. And we will also need to relate societies to one another. Exchanges with other societies may be functional substitutes for exchanges with a society's own natural environment, but this in turn has effects upon the other society's environment, so we shall have to be able to describe chains of effects across societies. Beyond that, we wish to be able to apply this term to human communities across history and around the globe.

In sociology, the term 'society' commonly refers to a social unit consisting of *a population within a certain territory, integrated by cultural commonalities*[9] as well as by political commonalities, such as shared procedures of decision-making, ways to enforce decisions, shared mutual responsibilities such as participatory duties, and a certain guarantee of care in the case of need (see, for example, Giddens 1989). While in sociology the idea of common governance (such as the modern nation state) is particularly important for the notion of society, cultural anthropology tends to stress the functional aspect of mutual interdependence and reproduction.[10]

For our purposes, this understanding of society makes sense. To conceive of society as a social unit functioning to reproduce a human population within a territory, guided by a specific culture, seems sufficiently abstract to be applied to different historical circumstances. It is still not very easy to determine the location of society in a hierarchy of social units (for example, household, community, state, federal state, the European Union or even the UN) or to determine whether American Indian kinship groups inhabiting a few settlements or different kinship groups meeting for winter camp, thereby providing an opportunity for culturally prescribed exogamous marriages, constitute a society. But perhaps it is not necessary to be precise at this point. What is more important here is to realize that society, according to this definition, links both elements that are symbolic, creations of social communication, and transmitted by communication between humans, subject to the rules of meaning and powerful for their meaning only ('culture'), *and* elements of a clearly natural origin and character, firmly subject to the rules of physics and biology ('population', 'territory'), in a way that has not yet been clearly specified.

If we turn our attention towards the 'idealist', 'platonic' tradition of the social sciences and look at the notion of society developed by the German sociologist Luhmann, we find that it is focused on communication, and on communication only. It appears devoid of any material, physical ingredients. Society, according to Luhmann, is the social system comprising all communication (Luhmann 1984; 1997). People, in the sense of physical persons, belong to society's environment, and so do all other material components such as habitat/territory, physical infrastructure or artefacts.

The power of Luhmann's notion of society lies in its consistent theoretical make-up. It is rooted in systems theory and rests upon a concept of the self-organization (or autopoiesis, that is, self-construction) of complex systems, and it invites us to think in terms of functional differences. Systems theory, we believe, provides a good basis for conceptualizing a world of meaning that is highly integrated internally and reproduces its boundaries vis-à-vis elements that do not 'fit'. This world of meaning and of communication is governed by completely different causalities (or inter-linkages) than the material world. The grammar of a sentence is neither subject to gravity, nor to genetic transmission, the presence or absence of a 'minus' will make a huge difference that cannot be influenced even by an atomic bomb, and a 'thank you' may change an entire situation for years, independent of any change in ambient temperature.

Why should anyone interested in society–nature interactions find Luhmann's characterization of social systems as systems of communication that are completely devoid of biophysical characteristics an important theoretical tool? A brief answer would be: because it allows the systematic conceptualization of those features of human society that distinguish it from biophysical or 'natural' systems. In addition, it frames them in terms of a theory of complex systems formally developed in biology (for example, Maturana and Varela 1975). This theory departs from self-referential operations that – if operationally closed – lead to system formation, that is, to the establishment of a boundary between a system and its environment. According to Luhmann, in social systems these self-referential operations are acts of communication that establish, in a recursive process, programmes and codes that distinguish between functionally differentiated subsystems and build up a high degree of internal complexity.[11]

On the other hand, the price of this 'purity' of focus on symbolic communication is impotence in relation to the material world. How can a purely symbolic system make a difference in terms of influencing biophysical objects? The answer is that it cannot do so. An outside agent is required to 'reach over' into material affairs. It seems obvious that this agent can be, and indeed has to be, the human as a hybrid of both realms, a sharer of symbolic understanding and thus a communicator, and a bodily creature with the ability to undertake physical action. Thus, while we accept Luhmann's conception of autopoietic systems of communication as a valuable tool, this cannot be all there is to social systems, it cannot be all that constitutes society. For the purpose at hand, this term must not be deprived of all material meaning; society must not be so exclusively self-referential that it cannot move so much as a chair.

It does seem practical to distinguish between 'society' and 'culture', and to employ both notions. Thus, adopting Luhmann's system of communication for culture[12] and allowing society to retain some material features appears to be the solution that most suits a socioecological purpose, enabling us to look upon society as a hybrid of the realms of culture, of meaning, of communication and of the material world.[13] Society, according to our understanding (for more detail, see Fischer-Kowalski and Weisz 1999), comprises both a cultural system, as a system of recurrent self-referential communication, and material components; a certain human population; and – this is the core of our understanding of society–nature interrelation – a physical infrastructure (buildings, machines, artefacts in use and animal livestock). It is via these biophysical components of society that culture interacts with nature: they may influence each other only indirectly and always according to their own rules. Conceived in this way, societies are not 'systems' in a strict sense of systems theory. They consist rather of a 'structural coupling' of a cultural system with material elements.

The very term 'social' seems to bear this hybrid character. That something is social implies the involvement of human beings, and human beings as persons can also reasonably only be thought of as an interlinking factor, a structural coupling between a symbolic (cultural, cognitive, mind) system that follows its own processes of self-creation and maintenance (having sampled much of its contents from society's cultural system), and a physical system, a body (that derives much of its functioning from genetic information sampled from the population's gene pool). A household may be conceived similarly: it is probable that a family culture exists that guides behaviour, physical human beings belong to it, and other physical elements sustain it and are sustained by human activities, such as a home, furniture, pets, a garden, a car and so on. Obviously, the organization of the household is maintained by communication, but not by communication alone. It also requires certain purely natural processes (such as walls remaining upright, roofs protecting from rainfall, heating to retain warmth inside, dogs stinking, plates breaking, and so on), and, as a go-between, human labour that moulds material elements to reproduce household functioning.

To conclude, we conceive of society as a structural coupling between a cultural system and material elements, among them, as its functional focus, human population. Society is simultaneously guided by two programmes, two types of software: a cultural software (that determines meaning and moulds intentions) and a natural software that determines material effectiveness. Society has this in common with all other *social* units. It is distinguished from them by particular features, in which we basically follow the traditions most common in sociology and cultural anthropology (see above).

Society in its entirety may thus not be regarded as a subsystem of an ecosystem, as it is often conceptualized within natural science-based sustainability research (Berkes and Folke 1998). On the other hand, we do not underestimate the relevance of biophysical aspects or conceptualize social systems as devoid of any material, biophysical ingredients as is the case in much of the research produced in the humanities, sociology and economics. According to our understanding, society comprises both a cultural system, as a system of recurrent self-referential communication, and material components; in other words, a certain human population as well as a physical infrastructure that includes buildings, machines, artefacts in use and animal livestock, which in their entirety may be defined as 'biophysical structures of society' (Fischer-Kowalski and Weisz 1999; Weisz et al. 2001, p. 121).

As Figure 1.1 shows, this notion of society allows an epistemological framework for the interaction of social and natural systems to be specified. It comprises a 'natural' or 'biophysical' sphere of causation governed by natural laws, and a 'cultural' or 'symbolic' sphere of causation reproduced by symbolic communication. These two spheres overlap, constituting what are termed here the 'biophysical structures of society'.[14] According to this concept, the process of interaction between nature and culture can only occur via these societal biophysical structures.

Socioecological Regimes, History and Transitions

If society is a hybrid, comprising an autopoietic cultural system, and material elements to which it is structurally coupled, then the very interactions between society and its material environment should be of decisive importance for the development of society itself. This is indeed the core hypothesis of Maurice Godelier, whose work has influenced our understanding of the interconnections between society, nature and history. Godelier formulates his core hypothesis in the introduction to *The Mental and the Material* thus: '*Human beings have a history because they transform nature*. It is indeed this capacity which defines them as human. Of all the forces which set them in movement and prompt them to invent new forms of society, the most profound is their ability to transform their relations *with* nature by transforming *nature itself*' (Godelier 1986, p. 1).

This way of looking at history relates to the Marxist tradition but transcends it in an ecological, or co-evolutionary direction. The classic reading of Marx leads to a discussion of changing 'modes of appropriation of nature' through the development of new means of production, that is,

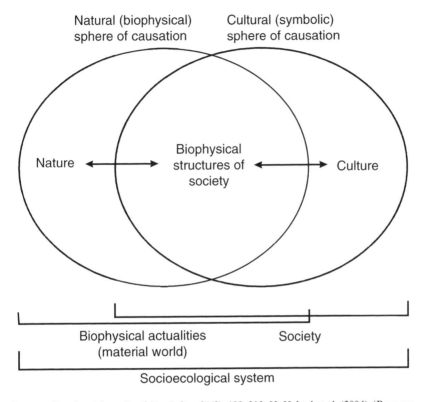

Natural (biophysical)
sphere of causation

Cultural (symbolic)
sphere of causation

Nature ⟷ Biophysical
structures of
society ⟷ Culture

Biophysical actualities
(material world)

Society

Socioecological system

Source: Reprinted from *Land Use Policy*, **21**(3), 199–213, H. Haberl et al. (2004), 'Progress towards sustainability? What the conceptual framework of material and energy flow accounting (MEFA) can offer', with permisison from Elsevier.

Figure 1.1   Socioecological systems as the overlap of a natural and a cultural sphere of causation

technology.[15] Godelier's reading stresses the fact that human appropriation of nature modifies nature and this modified nature in turn stimulates social change. Godelier thus deviates from the common social science approach by viewing nature as historically variable, not as static, and his core hypothesis attributes societies' historical dynamics to a feedback process from nature:

> The boundary between nature and culture, the distinction between the material and the mental, tend moreover to dissolve once we approach that part of nature which is directly subordinated to humanity – that is, produced or reproduced by it (domestic animals and plants, tools, weapons, clothes). Although external to us this nature is not external to culture, society or history. It is that part of nature

which is transformed by human action and thought. It is a reality which is simultaneously material and mental. It owes its existence to conscious human action
on nature. . . . This part of nature is appropriated, humanized, becomes society:
it is history inscribed in nature. (Godelier 1986, p. 4)

Thus, according to Godelier, the dynamic force in human history is not so
much the dialectics of 'means of production' and 'modes of production'
with nature as an external element, as something to be appropriated, but is
instead the very interaction between social and natural relations itself.

If we now look upon society as reproducing its population, we note that
it does so by interacting with natural systems, by organizing energetic and
material flows from and to its environment, by means of particular technologies and by transforming natural systems through labour and technology in specific ways to make them more useful for society's purposes. This
in turn triggers intended and unintended changes in the natural environment to which societies react. We regard this as a co-evolutionary process:
societies become structurally coupled with parts of their environment,
leading to a process where both mutually constrain each other's future evolutionary options.[16] The co-evolutionary process is then maintained by the
specific exchange relationship with the environment, by the particular way
a society interacts with certain natural systems. In this co-evolutionary
process, we can distinguish ideal-typical 'states', that is, patterns of
society–nature interactions that remain in a more or less dynamic equilibrium over long periods of time ('socioecological regimes'), and also periods
of transition.

In the most general terms, socioecological regimes in world history correspond to what many authors, using different terms, have addressed as
human modes of subsistence (Boyden 1992; Diamond 1997; Gellner 1988;
Sieferle 1997b). The transitions between these modes of subsistence are so
fundamental that they must often be called 'revolutions', namely, the
Neolithic Revolution (the transition from hunting-gathering to agrarian
society) and the Industrial Revolution (the transition from agrarian to
industrial society). These transitions provoke a number of questions. Why
did particular socioecological regimes not last forever, or, in other words,
why were they not sustainable? Why was there, for example, a transition
from hunting-gathering to the agrarian mode? And why, after roughly
10 000 years of agrarian societies, did a transition we call the Industrial
Revolution begin, leading to another mode, which is still so dynamic that
we find it hard to regard it at all as a defined mode of subsistence, that is,
as a socioecological regime with some dynamic stability? And how does a
possible future 'sustainable' socioecological regime that we might head for
stand in relation to all this? These are wide-ranging questions indeed, and

although we do not aspire to answer them in this book we do believe that they can provide the background needed to put the sustainability transition into perspective.

Looking back into history, we can also discuss the sustainability of previous alternative socioecological regimes. One of the most interesting discussions of this issue is provided by Sieferle (2003). According to Sieferle, hunters and gatherers sustain themselves through passive solar energy utilization, that is, their socioeconomic energy metabolism depends on the existing density of solar radiation and its transformation into plant biomass; they do not deliberately intervene in this transformation process.[17] Thus hunters and gatherers must more or less live on the resource density they find and as such, they can neither accumulate significant stocks of belongings nor seriously pollute their environment. The only sustainability threat they pose is in the form of overexploitation of key resources. For example, there is evidence that hunter-gatherers, although they probably consumed less than 0.01 per cent of the net primary production (NPP) of their habitat (Boyden 1992), contributed to the extinction of a significant part of the Pleistocene megafauna (that is, of animals with a body mass of over 10 kg, which were most suitable for hunting and therefore an important part of their resource base). Although the issue is highly controversial (Alroy 2001; Grayson et al. 2001), it does make a case for bringing the hunter-gatherer socioecological regime into debate, as far as sustainability is concerned. Nevertheless, this socioecological regime persisted for several hundred thousand years, certainly much longer than the prevalent industrial pattern, at least while it remains based on the use of fossil fuels and the large-scale use of exhaustible mineral resources.

Agrarian societies, to follow Sieferle, can be characterized by an energy regime of 'active solar energy utilization'. Their solar energy utilization is active insofar as they intervene into the solar energy transformation process by means of biotechnologies and by mechanical devices. The technological transformation of terrestrial ecosystems is of most importance: agrarian societies clear forests, create agro-ecosystems, breed new species and seek to extinguish other species. Their core strategy is the monopolization of area (and the corresponding solar radiation) for organisms with high utility for humans. Mechanical devices, on the other hand (such as the sailing boat or the watermill), transform solar energy occurring as wind or running water into a movement that can be utilized by humans.

Agrarian societies seem to have always struggled, with varying degrees of success, to maintain the delicate balance between population growth, agricultural technology, labour force needed to maintain the productivity of agro-ecosystems, and the maintenance of soil fertility (Netting 1981, 1993; Vasey 1992). Agrarian civilizations were always at risk, most often from

a combination of technological and political dependencies and the fluctuations of natural systems. Not only did the ancient Mesopotamians gradually degrade their soils by irrigation techniques, forcing peasants at first to
give up wheat cultivation for the more salt-tolerant barley and later to
abandon cultivation altogether, but also medieval peasants in the
Netherlands lost their fight against sand-dunes. Nevertheless, the agrarian
socioecological regime, in many parts of the world, persisted for several
thousand years and still persists today.

The presently dominant industrial socioecological regime dates back no
more than 300 years and is based upon the utilization of fossil fuels. Its sustainability seems limited not only by the limitations of its energy resource
base but also by the transformations it triggers globally in various life-
sustaining natural systems. Today, Global Change research provides ample
evidence that major human-induced changes can be found on any spatial
scale, from local to global, and are transforming the Earth system at an
increasing pace (Schellnhuber 1999; Turner et al. 1990). So this socioecological regime is bound to change as it erodes its natural base. In this situation, sustainability may involve guiding this transition within a corridor
of acceptable quality of life, for present and future human generations. The
MEFA framework as described below and referred to throughout this book
is our core device for analysing and understanding the metabolic exchange
relations between human societies and their natural environments, the feedbacks that transform both social and natural systems and the biophysical
limitations of the systems involved.

The MEFA Framework for Describing Society–Nature Interactions

Current approaches towards analysing the biophysical aspects of the earth
system (for example, Schellnhuber 1999; Schellnhuber and Wenzel 1998)
can be traced back to the work of ecologists (for example, Lotka 1925;
Lindemann 1942; Odum 1969) who conceptualized ecosystems using so-
called compartment models. In these models, ecosystems are analysed by
defining compartments, that is, black boxes transforming defined inputs
into outputs, according to some internal mechanisms and depending on
their own structure as well as on the state of all the other compartments of
the system. Ecosystem research proceeded by analysing the physical stocks
within and flows between the compartments and the mechanisms controlling these flows.[18]

Studying material and energy flows related to socioeconomic activities in
a similar way as 'socioeconomic metabolism' can be traced back at least as
far as this ecological research strategy (for reviews, see Fischer-Kowalski
1998; Martinez-Alier 1987). In the present context, the socioeconomic

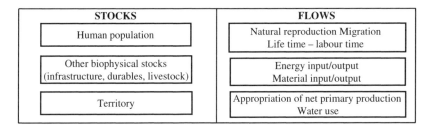

STOCKS	FLOWS
Human population	Natural reproduction Migration Life time – labour time
Other biophysical stocks (infrastructure, durables, livestock)	Energy input/output Material input/output
Territory	Appropriation of net primary production Water use

Figure 1.2   Biophysical dimensions of social systems

metabolism approach is attractive as it allows the biophysical structures of societies to be defined in a way that is compatible with the compartment models commonly used in systems ecology (Haberl 2001). That is, the metabolism approach allows one to analyse biophysical aspects of a society as if it were an ecosystem compartment, looking at its material stocks as well as the flows between the biophysical structures of society and the rest of the natural world. Generally speaking, the same concepts and methods can be used to deal with social and natural systems.

The stocks and flows listed in Figure 1.2 deliver a biophysical description of any society analogous to an ecosystem, and the interrelations between stocks and flows are – within a certain range – determined by natural processes. A description of these parameters is useful when analysing the interrelations and interdependencies between societies and their natural (and social) environments.

As a means of social science analysis, this is uncommon but hardly new. Anthropology, for example, has a long tradition of analysing the relation between simple societies and their natural environment by tracing energy flows (for example, White 1943; Rappaport 1971). Concerning complex modern societies, this approach can be traced back to the early 1970s (Ayres and Kneese 1969; Boulding 1973). It would not be as attractive, however, if it provided no more than a biophysical description. What it does in addition to producing biophysical descriptions is to establish a relation to the most powerful cultural system of modern society: the economy.[19] In particular, MEFA (material and energy flow accounting) seeks to analyse biophysical aspects of society in a way that is compatible with the most common and powerful tool for societal self-observation, the system of national accounts.

By means of this 'double compatibility', this approach establishes a link between socioeconomic variables on the one hand and biophysical patterns and processes on the other. Admittedly, MEFA remains an approach 'in the making'. It ultimately aims to provide a full systemic account of all the variables (and implied processes) listed in Figure 1.2 and their theoretical

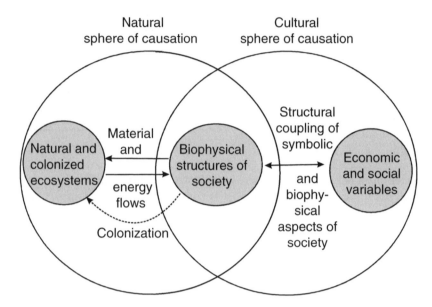

Source: Reprinted from *Land Use Policy*, **21**(3), 199–213, H. Haberl et al. (2004), 'Progress towards sustainability? What the conceptual framework of material and energy flow accounting (MEFA) can offer', with permisison from Elsevier.

Figure 1.3   *Biophysical structures of society simultaneously viewed as an ecosystem compartment and as structurally coupled with symbolic aspects of society*

integration. What exists so far, and is utilized in the following chapters, is a sociometabolic model describing material and energy flows (see Figure 1.3).

Another set of relations listed in Figure 1.3 centres around territory and land use as one of the most important socioeconomic pressures upon the environment and driving forces of global change (Meyer and Turner 1994; Vitousek 1992). We have theorized this set of relations under the heading of 'colonization of terrestrial ecosystems' (Fischer-Kowalski and Haberl 1998; Haberl et al. 2001; Krausmann et al. 2003).

While socioeconomic metabolism refers to the exchange of energy and matter between social and natural systems, colonization refers to society's deliberate interventions into natural systems in order to create and maintain a state of the natural system that renders it more useful socially (Fischer-Kowalski and Weisz 1999). Thus, colonization refers mainly to human labour and to the information, technologies and skills involved in making labour effective. Within the MEFA approach, this theoretical concept has become operational in describing land use. Socioeconomic land use can be related to changes in ecosystem patterns and processes. The impact of land use can be measured by comparing ecosystem patterns and processes that would be expected without human intervention with those observable in the presence of interventions. An example of this approach is the calculation of the 'human appropriation of net primary production', or HANPP (Vitousek et al. 1986).

Some other parts of the MEFA framework as outlined in Figures 1.2 and 1.3 remain very sketchy and will appear in this book only in rudimentary form. One of these still underdeveloped areas concerns population, its reproduction and time use. Yet another involves water availability and water use. MEFA does provide a consistent conceptual framework to deal with these issues and relate them to the other processes, but the provision of a comprehensive analytical and empirical body of research still lies some way ahead of us. In a similar vein, we have to admit to having made a largely descriptive use of statistical tools. Explicit causal models as well as uncertainty estimates shall constitute a next step in research terms.

1.4   THE DESIGN OF THIS BOOK

This book is an interim resumé of research carried out during the past decade by the team of the Institute of Social Ecology in Vienna, together with partners from all over the world. Within various contexts, more than 20 case studies were carried out that all follow the common conceptual and methodological framework referred to as MEFA in the preceding section. Regionally, these case studies range across Europe, India, Southeast Asia and Latin America. Temporally, some of them extend back over the past 170 years, others cover the past two or three decades, and still others refer to one or two points in time during the recent past. In terms of scale, they deal both with the local level (single communities) and the national level, and sometimes aggregate to regions.[20] Intellectual support in generating a coherent structure came from the International Human Dimensions of Global Environmental Change Programme (IHDP). Both the 'Industrial Transformation' (IHDP-IT) and the 'Land Use and Cover Change' (LUCC) sub-programmes endorsed our integrative research effort, in which we tried to build an umbrella over a number of cases, pursuing certain common research questions.

The point of departure for this research was the assumption that an eventual transition to a more sustainable socioecological regime would require as major a change as did the transition from agrarian to industrial society – a transition that is still ongoing in many parts of the world. Must it be assumed that the historical path will lead all regions of the world through a highly polluting and wasteful industrial stage eventually to some more sustainable regime? Would it not be worthwhile focusing attention on options of 'leapfrogging' certain developments? Since we believed that this should be considered, we wanted to attain a better understanding of what the agrarian socioecological regime entails and of what its potential for transformation is.

Another incentive for improving our understanding of the agrarian socio-ecological regime was the observation that several political demands from advocates of sustainability today provide reminders of the agrarian past, such as the appeal to return to biomass as the major source of energy, or the proposal that we should rely more on local resources and strengthen regional sustainability in the interests of self-sufficiency, or even the well-known 'factor 10/factor 4' debate on the reduction of materials use (Hinterberger and Schmidt-Bleek 1999; Weizsäcker, Lovins and Lovins 1997). These suggestions all raised questions of feasibility, in view of historical realities.

All these aspects inspired our research interest, which goes far beyond historical curiosity, to really understand what 'agrarian society' and the agrarian socioecological regime is like in terms of both the biophysical features of society and its transformative impact on nature. This required a substantial effort in terms of data generation. While a large body of literature of course exists concerning the socioeconomic structure of agrarian societies (both in terms of social, economic and technological history and of cultural anthropology) and a rich body of agro-historical and agro-ecological work, our specific focus was to be the society–nature interface. To explore this, we needed comparable quantitative data on many cases differing in important respects. We wished to provide answers to the following general questions:

1. Is there a 'characteristic metabolic profile' of agrarian societies, in terms of energy and materials use? Can a general claim be made on the basis of what has been observed in some cases that the energy and materials throughput of agrarian societies is approximately factor 3–5 smaller than that of industrial societies? Is such a metabolic profile connected to and dependent upon certain land-use patterns that can be described on a general level? Or rather is every agrarian society a unique case, depending on its specific natural conditions and socioeconomic history?

2. What happens when this socioecological regime starts to change? What are the major drivers of such a change? What new options open up, and which new risks appear? Which pressures upon the environment gain momentum and which pressures recede? What changes in natural systems can be observed during the transition?

3. How much does the course of the agrarian–industrial transition depend on the historical (world) context? What difference does it make

if a society is a worldwide pioneer of this transition (such as the United Kingdom), or is a comparative latecomer (such as the Austro-Hungarian Empire)? What, by comparison, does the transition look like when it takes place in a developing country, in a world context of a fully developed and dominant socioecological regime that is (post)industrial? Do common patterns exist with which all transitions comply, or must they be discussed on a strictly case-by-case basis?

4.   How does the interplay between different spatial scales and levels of society work? As nation states undergo a transition from the agrarian to the industrial mode, their peripheral rural communities may hardly be touched upon – but in the course of the process, they come under great pressure to undergo change themselves. So how do these scales and levels interact?

Clearly, we are unable to answer such wide-ranging questions in any definitive way on the basis of a few cases within any one book. Nonetheless, we still hope that we have been able to make a serious contribution towards finding these answers by our decision to narrow down the range of phenomena we observed and to invest a great deal of methodological effort in creating a common protocol for their measurement. While the conceptual guidelines (as described in the previous paragraph) were fairly clear from the start, the specific operationalizations still presented a challenge. Responding to this challenge implied, for example, creating methods of material and energy flow analysis that allowed the generation and modelling of these data from historical records (something quite different from generating them, as is usual, on the basis of contemporary economic statistics). It also implied creating methods to describe historical land use in a way that would be comparable to modern, GIS-based procedures. Finally, it required creating a full methodology to analyse the biophysical features of local rural communities, with the help of observations, on-site measurement and interviews, which would generate a database conceptually comparable to existing ones on the national level.

Somewhat more in the mainstream of material and energy flow analysis, but still a major methodological achievement, was the generation of MEFA databases for several developing countries that did not yet have any experience in this, and that had hardly any adequate economic statistics: this involved organizing extensive training for local experts, and supervising their work for extended time periods.[21] Thus, while the focus of this book is not primarily methodological, substantial parts of several chapters are devoted to explaining and discussing methodological issues.

As explained above, we have taken a comparative approach in attempting to find answers towards our 'grand' questions. We compare social systems, described according to their biophysical features, in transition from an agrarian to an industrial socioecological regime. This is performed in one of two ways: either long time series of observations exist across the period where the transition takes place (as in Chapters 2 to 4 and partially in Chapter 5), or we know from sources external to our primary research that we are dealing with societies in transition and may describe and/or compare their features at one specific point in time (as in Chapters 6 and 7 and some parts of Chapter 5). We treat each of the cases we have included in comparative analysis as a complex system with various strongly interdependent features. So although we have gathered and present quantitative data, we rarely compare results on single dimensions but try instead to take into account the complex mechanisms interlinking these data. This approach creates some difficulties in terms of presentation and we have yet to invent a standard procedure for this endeavour. Thus the reader will find that each chapter seeks its own way to respond to this challenge. Furthermore, the cases presented are not only complex cases in their own right, but they are ordered on different sides of important distinctions (see Figure 1.4).

While each case in itself represents a transition from the agrarian to the industrial socioecological regime, it may be characterized according to its position on two relevant axes. One axis refers to spatial scale (or level of organization) and distinguishes between local and national cases. Chapter 5 is particularly devoted to this issue, as it compares the long historical time

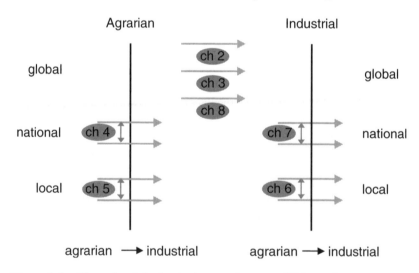

Figure 1.4   The role of the book chapters in exemplifying relevant distinctions

series data on the national level discussed in Chapter 2 with data on various local communities between the years 1830 and 2000. This chapter is intended to make a specific contribution to the question of how different levels in transitions interrelate. Other chapters (such as Chapters 6 and 7) also contribute to issues of spatial scale, although their main focus is different.

The other axis characterizes the world context. During the whole time period we are dealing with (from about 1600 to 2000) the 'great transformation' takes place, and in Chapters 2 and 3 we try to capture the overall process. While Chapter 2 attempts a grand view as to what this transformation implies in terms of socioecological regime change, Chapter 3 deals with some important implications of this for global environmental change, in particular with the impacts upon the carbon cycle. As this transformation goes on and gradually spreads (both regionally and in depth), the context for 'newcomers' changes. It should make a significant difference whether a country's transition from the agrarian to the industrial socioecological regime occurs as a pioneer case, or if it can make use of technologies and organizational experiences from elsewhere, but also has to reach an accommodation with the established economic power structures created and dominated by forerunners. This question is dealt with for specific cases in Chapters 5, 6 and 7, while Chapter 8 then systematically compares the paths to industrialization of the (pioneering) United Kingdom and the (latecoming) Austro-Hungarian Monarchy, and contrasts them with ongoing changes in contemporary developing countries. There exist several other relevant differences between those former empires, of course, such as access to overseas colonies or the transport and trade conditions of an island versus a geographical location in alpine regions of a large continent, and each developing country's history again makes for a particular case that we represent only insufficiently in this book. As such, our explanations are bound to be far from exhaustive; the interesting question, however, is whether we can establish similarities across those differences, that may legitimately be attributed to the respective socioecological regime.

Chapters 6 and 7 deal with social systems that are still in transition from the agrarian to the industrial regime themselves, but in a world context that is already completely dominated by the industrial socioecological regime. They all fall under the label of 'developing countries' yet are highly differentiated among each other, ranging from Brazil to some of the poorest countries of the world such as Laos. For the local case studies in Chapter 6, it is hard to tell in what way this world context matters but the chapter's analysis addresses the question of what difference the national (usually also transitional) context makes. The comparisons in Chapter 7, on the other hand, deal with the national scale and draw their main explanatory power from relating this to the global context and analysing the specific role a country occupies within an international division of labour as a determinant of transitional pathways.

Finally, Chapter 8 seeks to knit together those strands of insight generated by the previous chapters towards formulating preliminary answers to our major questions. It also critically reviews the premises of the whole enterprise: whether it makes sense to use the sociometabolic transition from the agrarian to the industrial regime as a key to unlocking better understanding of possible transitions towards sustainability.

The central theme underlying this book is the notion that most, if not all, global sustainability issues have to do with the fact that about two-thirds, if not three-quarters of the world population are currently in the midst of a rapid transition from agrarian society to the industrial regime. This transition is fundamentally changing societal organization, economic structures, patterns of resource use and so on, thereby probing the limitations of the planet Earth in many ways, among others by using up exhaustible resources, altering global biogeochemical cycles, depleting biological diversity and degrading the Earth's ecosystems (Millennium Ecosystem Assessment 2005). All of this is triggering a process of global environmental change comparable to past transitions in the Earth system such as those between Ice Ages and warm interglacial periods (Schellnhuber et al. 2004; Steffen et al. 2004). What seems clear is that neither the Earth's mineral resources, nor the adaptive capacity of the biosphere will suffice to allow the transition of six or even nine billion people (Lutz et al. 2004) to the current pattern of industrial resource use: humanity's use of natural resources already exceeds the regenerative capacity of the biosphere by at least 20 per cent (Wackernagel et al. 2002).

We are sceptical about the notion that a clearcut set of measures could be identified here and now that would bring about a transition towards sustainability. Rather we subscribe to the view that a better understanding of this transition process may help us to identify ways to intervene in ongoing transitions in a useful way. Overall, this book is intended to contribute to this analytic endeavour.

NOTES

1. During this period, water use and pig population, for example, increased ninefold, the world economy 14-fold, energy consumption and CO^2 emissions 17-fold, world trade 22-fold, marine fish catch 35-fold, industrial output 40-fold and, finally, world freight transport by a factor of 135 (McNeill 2000).
2. In mathematical terms, growth or modernization processes can be adequately described by correlations and econometric models, whereas for transition processes

different models are required that can capture the discontinuity of interrelations (Ayres 2000).
3. At least in principle, this competition could be resolved mathematically by finding the model, or the group of models, with a better fit. It is nonetheless more interesting to compare the intellectual service each of the theories provides.
4. There is an obvious similarity between the 'stabilization' phase and the highly connected, stable phase and between the 'pioneer'-dominated third transient phase and the 'acceleration' phase. Whether or not it is helpful for the analysis of socioecological systems to break up the 'take-off' phase in two phases, one of release and one of reorganization, may depend on the circumstances.
5. For a careful analysis of types of evolutionary theories in biology and the social sciences see Weisz (2002).
6. See also the book series *Routledge Studies of Societies in Transition*.
7. It is interesting to note that in the *Communist Manifesto* (Marx and Engels [1847] 1998) the advent of socialism appears as inevitable as the advent of capitalism before it. In his later writings, Marx became more cautious about imputing such mechanics to history.
8. Analytically, these approaches owe much to the research tradition on patterns of technological change in the past, such as Grübler (1994); Nakicenovic (1990); Perez (1983).
9. Such as a common language, a system of legislation, a currency and a certain minimum of shared everyday customs and habits.
10. As in the textbook definition by Harris, according to which society is an 'organized group of people who share a habitat and who depend on each other for their survival and well-being'. 'Each society has an overall culture', he adds, which, however, need not be uniform for all members (Harris 1987, p. 10).
11. Luhmann has spent more than two decades developing his theory in a very thorough and consistent way (Luhmann 1984; 1997). As a social scientist, it is very helpful to be able to draw on such a well-integrated body of groundwork. English translations of some of Luhmann's major works have only recently become available (see, for example, Luhmann 1995).
12. The terminology is close to Rolf Peter Sieferle's way of looking at culture (Sieferle 1997a), but rather far from the traditions of cultural anthropology and of most sociological writers.
13. The term 'hybrid' as a scientific term originates from biology, where it refers to the 'offspring of a cross between two different strains, varieties, races, or species' (Walker 1989). Within the social sciences, the French science critic Bruno Latour (1998) in particular made this term central to his theoretical approach. According to Latour, hybrids signify human creations that, inseparably, are both creatures of culture and creatures of nature. Latour argues that pre-modern thinking had assumed a certain unity of the social and the natural, the human and the non-human, nature and culture, the subject and the object. The idea of separating nature and society belongs to the very constitution of modernity (Latour 1998, p. 22). But while modern thinking, modern science, strips its natural objects of all communicative capacity and political relevance on the one hand and frees society of natural (or transcendent) determination on the other, this results in continuously creating mixed creatures (hybrids, sometimes also called 'monsters') that exert a powerful influence upon the destinies of both society and nature. The unconsciousness with which this is performed, the lack of conceptual tools to deal with these hybrids and the built-in ambiguity of the distinctions, on one hand makes modern thinking invincible (p. 53) while on the other it fills the world with potentially dangerous 'quasi-objects'.
14. This concept is, to some extent, similar to that proposed by Schellnhuber to define the Earth system as consisting of two main components, N and H, where N (the natural system) is assumed to consist of components such as the atmosphere, the biosphere, the cryosphere and so on, and H (denoted by Schellnhuber as 'the human factor') consists of the so-called 'physical subcomponent' or 'anthroposphere' (in the words of Schellnhuber 1999 [C20]: 'the aggregate of all human lives, actions and products') and a 'metaphysical' sub-component roughly comparable to the notion of 'culture' as employed in Figure 1.1 of this chapter. However, Schellnhuber's concept lacks an understanding of human

society – both in its physical and its 'metaphysical', or cultural aspects – as a complex autopoietic system.

15. Foster (2000) makes a very convincing attempt to demonstrate that Marx was indeed aware of the ecological consequences of changing means of production and that he did see natural conditions as changing and thereby influencing society. However, this line of Marx's thinking seems to have been almost entirely restricted to agriculture and did not shape his overall model of historical transitions.

16. See Goudsblom's 'principle of paired increases in control and dependency' (Goudsblom, Jones and Mennell 1996, p. 25).

17. They already differ from all other mammals, however, through the utilization of fire. Burning wood is a method of transcending contemporaneous energy flows by mobilizing stocks; on the other hand, controlled burning of dry vegetation alters the landcover and makes regions more accessible for hunting (Goudsblom 1992).

18. Current research into the global carbon cycle – an important aspect of earth system analysis – still proceeds exactly in this way (for example, Houghton 1995; Houghton and Skole 1990).

19. It may appear unusual to call the economy a 'cultural system'. To justify this definition, we argue as follows. Our point of departure is the distinction between a 'natural' and a 'cultural' sphere of causation (see Figure 1.1). If we follow Luhmann's specification of subsystems of society, the economy deals in the medium of money, a symbolic device representing value. While classical 'political economy' theorists (Smith, Ricardo, Marx) still formulated their theories in a 'hybrid' fashion, referring equally to physical and monetary units, 20th-century economics increasingly liberated itself from a physical reference. This change seems not to have taken place only on the level of scientific discourse, but also on the level of the corresponding 'reality', where financial phenomena are increasingly decoupled from any physical process.

20. The primary empirical material of these case studies was generated by projects financed by the European Commission (especially the FP5-Incodev programme), the statistical office of the European Union, Eurostat, the Austrian Federal Ministry of Science's social science programmes (Nicobars: Indian and Cultural Studies), the Breuninger Foundation and the Austrian National Science Fund. Crucial funding came from the Austrian Ministry of Science's research programme on 'Cultural Landscapes' – this funding allowed us to go beyond single case studies and engage in comparative analysis.

21. Our philosophy, in accordance with the respective EU-financed research programme, was not just to generate data on developing countries but to stimulate the growth of a local capacity to do so in terms of fostering expertise. At the same time, we conceived incentives for them to look comparatively at each other and learn from each other, and not to fix their attention upon us and the other European research teams involved. In selected cases, our efforts in capacity building went as far as inviting collaborating researchers to undertake their PhD research in Vienna and then return as senior scientists to their countries. We thank IHDP-IT for supporting this with recommendations for European START scholarships, as well as the ASEAN bank for its help. We also thank IHDP-IT for endorsing, and sometimes supporting financially, regional workshops (such as those in Manaus in 2001, in Ho Chi Minh city in 2002 and in Rio de Janeiro in 2003).

REFERENCES

Alroy, John (2001), 'A Multispecies Overkill Simulation of the End-Pleistocene Megafaunal Mass Extinction', *Science*, **292**, 1893–6.
Ayres, Robert U. (2000), 'On Forecasting Discontinuities', *Technological Forecasting and Social Change*, **65**(1), 81–97.
Ayres, Robert U. and Allen V. Kneese (1969), 'Production, Consumption and Externalities', *American Economic Review*, **59**(3), 282–97.

Berkes, Fikret and Carl Folke (1998), 'Linking Social and Ecological Systems for Resilience and Sustainability', in Fikret Berkes and Carl Folke (eds), *Linking Social and Ecological Systems. Management Practices and Social Mechanisms for Building Resilience*, Cambridge: Cambridge University Press, pp. 1–26.

Berkhout, Frans, Adrian Smith and Andy Stirling (2003), *Socio-technical Regimes and Transition Contexts*, Brighton: SPRU, University of Sussex.

Boulding, Kenneth E. (1973), 'The Economics of the Coming Spaceship Earth', in Herman E. Daly (ed.), *Towards a Steady State Economy*, San Francisco, CA: Freeman, pp. 3–14.

Boyden, Stephen V. (1992), *Biohistory: The Interplay Between Human Society and the Biosphere – Past and Present*, Paris, Casterton Hall, Park Ridge, New Jersey: UNESCO and Parthenon Publishing Group.

Brand, Karl-Werner (1997), 'Probleme und Potentiale einer Neubestimmung des Projekts der Moderne unter dem Leitbild "Nachhaltige Entwicklung". Zur Einführung', in Karl-Werner Brand (ed.), *Nachhaltige Entwicklung, Eine Herausforderung an die Soziologie*, Opladen: Leske & Budrich, pp. 9–32.

Diamond, Jared M. (1997), *Guns, Germs, and Steel: The Fates of Human Societies*, New York, London: W.W. Norton & Company.

Fischer-Kowalski, Marina (1998), 'Society's Metabolism. The Intellectual History of Material Flow Analysis, Part I: 1860–1970', *Journal of Industrial Ecology*, **2**(1), 61–78.

Fischer-Kowalski, Marina and Helmut Haberl (1998), 'Sustainable Development: Socio-Economic Metabolism and Colonization of Nature', *International Social Science Journal*, **158**(4), 573–87.

Fischer-Kowalski, Marina and Helga Weisz (1999), 'Society as Hybrid Between Material and Symbolic Realms. Toward a Theoretical Framework of Society–Nature Interrelation', *Advances in Human Ecology*, **8**, 215–51.

Foster, John B. (2000), *Marx's Ecology. Materialism and Nature*, New York: Monthly Review Press.

Geels, Frank W. (2004), 'Understanding Technological Transitions: A Critical Literature Review and a Pragmatic Conceptual Synthesis', in B. Elzen, F.W. Geels and K. Green (eds), *System Innovation and the Transition to Sustainability: Theory, Evidence and Policy*, Cheltenham, UK and Northampton, MA, USA: Edward Elgar.

Geels, Frank W. (2006), 'Multi-level Perspective on System Innovation: Relevance for Industrial Transformation', in Xander Olsthorne and Anna Wieczorek (eds), *Understanding Industrial Transformation*, Dordrecht: Springer, pp. 163–86.

Gellner, Ernest (1988), *Plough, Sword and Book*, London: Collins Harvill.

Giddens, Anthony (1989), *Sociology*, Cambridge: Polity Press.

Godelier, Maurice (1986), *The Mental and the Material: Thought, Economy and Society*, London: Blackwell Verso.

Goudsblom, Johan (1992), *Fire and Civilization*, London: Penguin Books.

Goudsblom, Johan, Eric Jones and Stephen Mennell (1996), *The Course of Human History. Economic Growth, Social Process, and Civilization*, Armonk, NY and London: M.E. Sharpe.

Grayson, Donald K., John Alroy, Richard Slaughter and Joseph Skulan (2001), 'Did Human Hunting Cause Mass Extinction?', *Science*, **294**, 1459–62.

Grübler, Arnulf (1994), 'Industrialization as a Historical Phenomenon', in Robert H. Socolow et al. (eds), *Industrial Ecology and Global Change*, Cambridge, MA: Cambridge University Press, pp. 43–67.

Gunderson, Lance and C.S. Holling (2002), *Panarchy. Understanding Transformations in Human and Natural Systems*, Washington, DC: Island Press.

Haberl, Helmut (2001), 'The Energetic Metabolism of Societies, Part I: Accounting Concepts', *Journal of Industrial Ecology*, **5**(1), 11–33.

Haberl, Helmut, Karl-Heinz Erb, Fridolin Krausmann, Wolfgang Loibl, Niels B. Schulz and Helga Weisz (2001), 'Changes in Ecosystem Processes Induced by Land Use: Human Appropriation of Net Primary Production and its Influence on Standing Crop in Austria', *Global Biogeochemical Cycles*, **15**(4), 929–42.

Haberl, H. et al. (2004), Progress Towards Sustainability? What the Conceptual Framework of Material and Energy Flow Accounting (MEFA) Can Offer?, *Land Use Policy*, **21**(3), 199–213.

Harris, Marvin (1987), *Cultural Anthropology*, New York: Harper & Collins.

Hinterberger, Friedrich and Friedrich Schmidt-Bleek (1999), 'Dematerialization, MIPS and Factor 10, Physical Sustainability Indicators as a Social Device', *Ecological Economics*, **29**, 53–6.

Holling, C.S. (1973), 'Resilience and Stability of Ecological Systems', *Annual Review of Ecological Systematics*, **4**, 1–24.

Houghton, Richard A. (1995), 'Land-use Change and the Carbon Cycle', *Global Change Biology*, **1**, 275–87.

Houghton, Richard A. and David L. Skole (1990), 'Carbon', in Billie L.I. Turner et al. (eds), *The Earth as Transformed by Human Action, Global and Regional Changes in the Biosphere over the Past 300 Years*, Cambridge: Cambridge University Press, pp. 393–408.

Kemp, René and Derk Loorbach (2006), 'Transition Management: A Reflexive Governance Approach', in J.P. Voss, D. Bauknecht and R. Kemp, *Reflexive Governance for Sustainable Development*, Cheltenham, UK and Northampton, MA, USA: Edward Elgar, pp. 103–30.

Krausmann, Fridolin, Helmut Haberl, Niels B. Schulz, Karl-Heinz Erb, Ekkehard Darge and Veronika Gaube (2003), 'Land-use Change and Socioeconomic Metabolism in Austria. Part I: Driving Forces of Land-use Change: 1950–1995', *Land Use Policy*, **20**(1), 1–20.

Latour, Bruno (1998), *Wir sind nie modern gewesen. Versuch einer symmetrischen Anthropologie*, Frankfurt a.M.: Fischer Taschenbuch Verlag.

Lindemann, R.L. (1942), 'The Trophic-Dynamic Aspect of Ecology', *Ecology*, **23**(4), 399–418.

Lotka, Alfred J. (1925), *Elements of Physical Biology*, Baltimore, MD: Williams & Wilkins Company.

Luhmann, Niklas (1984), *Soziale Systeme. Grundriss einer allgemeinen Theorie*, Frankfurt a.M.: Suhrkamp.

Luhmann, Niklas (1995), *Social Systems*, Stanford, MD: Stanford University Press.

Luhmann, Niklas (1997), *Die Gesellschaft der Gesellschaft*, Frankfurt a.M.: Suhrkamp.

Lutz, Wolfgang, Warren C. Sanderson and Sergei Scherbov (2004), *The End of World Population Growth in the 21st Century. New Challenges for Human Capital Formation and Sustainable Development*, London, Sterling, VA: Earthscan.

Martens, Pim and Jan Rotmans (2002), *Transitions in a Globalising World*, Lisse, The Netherlands: Swets & Zeitlinger Publishers.

Martinez-Alier, Joan (1987), *Ecological Economics. Energy, Environment and Society*, Oxford: Blackwell.

Marx, Karl and Friedrich Engels [1847] (1998), *The Communist Manifesto*, Oxford: OUP Oxford Paperbacks.

Maturana, Humberto R. and Francisco G. Varela (1975), *Autopoietic Systems. A Characterization of the Living Organization*, Urbana-Champaign, IL: University of Illinois Press.

May, Robert M. (1977), 'Thresholds and Breakpoints in Ecosystems with a Multiplicity of Stable States', *Nature*, **269**, 471–7.

McNeill, John R. (2000), *Something New Under the Sun – An Environmental History of the Twentieth Century*, London: Allen Lane.

Meadows, Dennis L., Donella H. Meadows and Jorgen Randers (1972), *The Limits to Growth*, New York: Universe Books.

Meyer, William B. and Billie L.I. Turner (eds) (1994), *Changes in Land Use and Land Cover, A Global Perspective*, Cambridge: Cambridge University Press.

Millennium Ecosystem Assessment (2005), *Ecosystems and Human Well-being – Synthesis*, Washington, DC: Island Press.

Nakicenovic, Nebojsa (1990), 'Dynamics of Change and Long Waves', in R. Vasko et al. (eds), *Life Cycles and Long Waves. Lecture Notes in Economics and Mathematical Systems*, Berlin: Springer Verlag, pp. 147–92.

Netting, Robert M. (1981), *Balancing on an Alp. Ecological Change and Continuity in a Swiss Mountain Community*, London, New York, New Rochelle, Melbourne, Sydney: Cambridge University Press.

Netting, Robert M. (1993), *Smallholders, Householders. Farm Families and the Ecology of Intensive, Sustainable Agriculture*, Stanford, CA: Stanford University Press.

Odum, Eugene P. (1969), 'The Strategy of Ecosystem Development. An Understanding of Ecological Succession Provides a Basis for Resolving Man's Conflict With Nature', *Science*, **164**, 262–70.

Perez, C. (1983), 'Structural Change and the Assimiliation of New Technologies in the Economic and Social System', *Futures*, **15**(4), 357–75.

Polanyi, Karl (1944), *The Great Transformation*, New York: Farrar & Rinehart.

Rappaport, Roy A. (1971), 'The Flow of Energy in an Agricultural Society', *Scientific American*, **224**(3), 117–32.

Rip, A. and René Kemp (1998), 'Technological Change', in Steven Rayner and E. Malone (eds), *Human Choices and Climate Change 2*, Columbus, OH: Battelle.

Rotmans, Jan, René Kemp and Marjolein van Asselt (2001), 'More Evolution than Revolution: Transition Management in Public Policy', *Foresight*, **3**(1), 15–31.

Scheffer, Marten, Steve Carpenter, Jonathan A. Foley, Carl Folke and Brian Walker (2001), 'Catastrophic Shifts in Ecosystems', *Nature*, **413**, 591–6.

Schellnhuber, Hans J. (1999), ' "Earth System" Analysis and the Second Copernican Revolution', *Nature*, **402** (Suppl.), C19–C23.

Schellnhuber, Hans J. and V. Wenzel (1998), *Earth System Analysis. Integrating Science for Sustainability*, Berlin, Heidelberg: Springer.

Schellnhuber, Hans-Joachim, Paul J. Crutzen, William C. Clark, Martin Claussen and Hermann Held (2004), *Earth System Analysis for Sustainability*, Cambridge, MA, London, UK: Report of the 91st Dahlem Workshop, The MIT Press.

Schumpeter, J.A. (1950), *Capitalism, Socialism and Democracy*, New York: Harper & Row.

Sieferle, Rolf P. (1997a), 'Kulturelle Evolution des Gesellschaft-Natur-Verhältnisses', in Marina Fischer-Kowalski et al. (eds), *Gesellschaftlicher Stoffwechsel und Kolonisierung von Natur. Ein Versuch in Sozialer Ökologie*, Amsterdam: Gordon & Breach Fakultas, pp. 37–53.

Sieferle, Rolf P. (1997b), *Rückblick auf die Natur: Eine Geschichte des Menschen und seiner Umwelt*, München: Luchterhand.

Sieferle, Rolf P. (2003), 'Sustainability in a World History Perspective', in Brigitta Benzing (ed.), *Exploitation and Overexploitation in Societies Past and Present. IUAES-Intercongress 2001 Goettingen*, Münster: LIT Publishing House, pp. 123–42.

Steffen, W., A. Sanderson, P.D. Tyson, J. Jäger, P.A. Matson, B. Moore III, F. Oldfield, K. Richardson, H.J. Schellnhuber, B.L. Turner II and R.J. Wasson (2004), *Global Change and the Earth System. A Planet Under Pressure*, Berlin: Springer.

Turner, Billie L.I., William C. Clark, Robert W. Kates, John F. Richards, Jessica T. Mathews and William B. Meyer (1990), *The Earth as Transformed by Human Action: Global and Regional Changes in the Biosphere over the Past 300 Years*, Cambridge: Cambridge University Press.

Varela, Francisco G., Humberto R. Maturana and R. Uribe (1974), 'Autopoiesis: The Organization of Living Systems, its Characterization and a Model', *Biosystems*, **5**, 187–96.

Vasey, Daniel E. (1992), *An Ecological History of Agriculture, 10 000 B.C.–A.D. 10 000*, Ames: Iowa State University Press.

Vitousek, Peter M. (1992), 'Global Environmental Change: An Introduction', *Annual Review of Ecology and Systematics*, **23**, 1–14.

Vitousek, Peter M., Paul R. Ehrlich, Anne H. Ehrlich and Pamela A. Matson (1986), 'Human Appropriation of the Products of Photosynthesis', *BioScience*, **36**(6), 363–73.

Wackernagel, Mathis, Niels B. Schulz, Diana Deumling, Alejandro C. Linares, Martin Jenkins, Valerie Kapos, Chad Monfreda, Jonathan Loh, Norman Myers, Richard B. Norgaard and Jorgen Randers (2002), 'Tracking the Ecological Overshoot of the Human Economy', *Proceedings of the National Academy of Science*, **99**(14), 9266–71.

Walker, Peter M.B. (ed.) (1989), *Chambers Biology Dictionary*, Cambridge: Chambers.

Weisz, Helga (2002), 'Gesellschaft-Natur Koevolution: Bedingungen der Möglichkeit nachhaltiger Entwicklung', Dissertation, Humbold Universität, Berlin.

Weisz, Helga, Marina Fischer-Kowalski, Clemens M. Grünbühel, Helmut Haberl, Fridolin Krausmann and Verena Winiwarter (2001), 'Global Environmental Change and Historical Transitions', *Innovation – The European Journal of Social Sciences*, **14**(2), 117–42.

Weizsäcker, Ernst U.v., Amory B. Lovins and Hunter L. Lovins (1997), *Factor Four – Doubling Wealth, Halving Resource Use. The New Report to the Club of Rome*, London: Earthscan.

White, Leslie A. (1943), 'Energy and the Evolution of Culture', *American Anthropologist*, **45**(3), 335–56.

Willke, Helmut (1996), *Systemtheorie II: Interventionstheorie*, Stuttgart: Lucius & Lucius.

2. Land-use change and socioeconomic metabolism: a macro view of Austria 1830–2000

Fridolin Krausmann and Helmut Haberl

2.1 INTRODUCTION

Transitions from agrarian to industrial society are characterized by fundamental rearrangements in societal organization, in the economy (Gellner 1989; Polanyi 1971; Wrigley 1988) and in society–nature interaction (McNeill 2000; Turner et al. 1990). Focusing on this latter aspect, this chapter analyses the transition from agrarian to industrial society in Austria, one of the relatively late developers in terms of European industrialization. The analysis of aggregate changes on the national scale presented in this chapter is complemented by local case studies and a discussion of exchange relations between different locales in Chapter 5.

When we say 'Austria', we refer to the country with its present national boundaries. The socioeconomic system of Austria has only existed as a political and administrative unit since 1918. Present-day Austria extended before the end of World War I across several provinces of the Austro-Hungarian Empire, one of the largest empires in Europe. Five of the Austro-Hungarian Empire's provinces (*Kronländer*) were approximately identical to modern Austrian provinces (*Bundesländer*). These five provinces (counted today as six with the inclusion of the city of Vienna, which is now a separate province in administrative terms) account for almost 60 per cent of Austrian territory today. Two other provinces, Tyrol and Styria (35 per cent of Austria's current territory), covered a considerably larger territory before 1918. Therefore, historical data needed to be selected and adjusted to fit into a reasonably consistent time series.[1]

We will demonstrate how transitions from agrarian to industrial societies, and above all the changes in society–nature interaction they entail, can be assessed empirically by accounting for changes in socioeconomic metabolism and land use over long periods of time. Our analysis focuses on energy flows because they are considered a core element of socioecological regimes (Fischer-Kowalski 2003) and because changes in the use of energy resources – an important aspect of technological change (Grübler 1998) – were major drivers of industrialization.

We claim that it is possible to distinguish clearly between two broad types of socioecological regimes, an agrarian and an industrial, characterized by fundamental differences in their energy systems. The transition between the two is one of the most fundamental processes occurring during industrialization (Pomeranz 2000; Sieferle 1997, 2001; Smil 1992). Of course, both the agrarian and the industrial pattern are subject to considerable variation in both time and space and some of this variability can be seen in the case studies presented in other chapters of this volume. Nevertheless, we believe that society's energetic metabolism changes during the transition from agrarian to industrial society to an extent that justifies the definition of two distinct types of energy systems, agrarian and industrial.

The energy system of the agrarian regime is based mainly on the extraction of biomass from the local environment through agriculture and forestry. This type of energy system has been termed 'controlled solar energy system' (Sieferle 1997) because – in contrast to hunter-gatherers not discussed further here (see Boyden 1992) – most biomass comes from agro-ecosystems or managed forests, that is, from ecosystems intentionally transformed or 'colonized' by human society (Fischer-Kowalski and Haberl 1997).

Due to the reliance on an area-dependent, renewable primary energy source, the agrarian socioecological regime is subject to strict limitations with respect to physical growth and spatial differentiation. As discussed in Chapter 8, among the limiting factors are the comparatively low energy yield per unit of land area, the low energy density of biomass and the high energy costs of transport under the conditions of the agrarian regime. The main bottleneck for growth appears to be the problem of maintaining soil fertility without fossil-fuel-derived energy and nutrient subsidies. This only allows for comparatively low energy yields per unit of land area and requires a positive energetic return of agriculture per unit of energy invested (Pimentel and Pimentel 1996). Together, these factors prohibit physical growth of socioeconomic systems beyond a certain threshold[2] and determine the basic characteristics of the metabolic profile of the agrarian regime. As discussed in detail in Chapter 5, it is estimated that the controlled solar energy system under temperate climatic conditions allows for an energy yield of 30–50 gigajoules per hectare (GJ/ha) and maximum population densities of 30–40 cap/km^2. Material use may range from 5–6 tons per capita, biomass accounting for more than 75 per cent of total material use.

Practically all the limitations of the controlled solar energy system have been overcome by the transformation of the energy system during the process of industrialization. The energy system of industrial socioecological regimes is based to a large extent on area-independent fuels, above all, fossil fuels, and has thus been called the 'fossil energy system' (Sieferle 1997).[3] While agrarian societies gain most materials and almost all of their energy through agriculture and forestry, industrial societies rely heavily on non-renewable resources – minerals, fossil fuels, metals and so on – for the supply of energy and materials. Biomass typically accounts for less than 30 per cent of energy and 15 per cent of material use in industrial societies. Moreover, due to the emergence of far-reaching, efficient means of transportation (for which fossil fuels were a prerequisite), they can use energy and materials from distant places. The transformation of the energy system abolished the historical limits for physical growth and triggered an unprecedented surge in material and energy use. Socioeconomic turnover of energy and materials increased by one and more orders of magnitude and large-scale spatial differentiation and urban concentration became possible. Thus, the transformation of the socioeconomic energy system has to be considered as the core biophysical process with respect to the transformation of the socioecological regime and industrialization.

Fossil fuels allowed for a de-linking of energy provision and area. The transition from the agrarian to the industrial socioecological regime thus results in fundamental changes in land use, which usually also entail considerable changes in land cover. This will be demonstrated in this chapter by looking at changes in ecological energy flows induced by land use. Through land use, humans completely alter the energy flows of the ecosystems they depend on and compete with most other species for trophic energy, resulting in a process called 'human appropriation of net primary production' or HANPP (Haberl 1997; Vitousek et al. 1986). HANPP is defined as the difference between the NPP (net primary production) of potential vegetation, that is, the vegetation that is assumed to prevail in the absence of human intervention (Tüxen 1956), and the amount of NPP remaining in ecosystems under current land-use practices.

This chapter explores the specific course of the transition and discusses the metabolic patterns of different socioecological regimes as they appear in the Austrian case. It presents an analysis of the transformation of the energy system from the early 19th century until today with respect to energy provision and energy use. In doing so, it puts a special emphasis on the relation between the energy system and land use and the fundamental restructuring of this relation during the transition process. The discussion of these trends is based on a unique dataset for Austria for the time period from 1830 to 2000 (Krausmann 2001, 2003, 2004; Krausmann and Haberl 2002; Krausmann and Haberl et al. 2003). For this period, complete time series are available for many parameters and it encompasses the full time period in which Austria's transition from an agrarian to an industrial society took place. Austria can be considered to have been an advanced agrarian society in 1830. The energy system relied almost exclusively on biomass; hydropower and coal contributed less than 1 per cent to the energetic metabolism.[4] In 1995, the corresponding figure was nearly 70 per cent (Krausmann and Haberl 2002). During this period, the share of the agricultural population (that is, farmers and their families) declined from 75 per cent to 5 per cent. Total GDP rose by a factor of 28 and per-capita GDP by a factor of 12, while the contribution of agriculture and forestry to GDP dropped to 1.4 per cent in 2000. Austria's population increased by a factor of 2.3 from roughly 3.6 to 8.1 million. These changes are summarized in Figure 2.1.

While this chapter analyses the general trends at the national level, Chapter 5 provides a complementary discussion of empirical evidence at a regional level, looking at two rural villages and Austria's capital (and by far largest city) Vienna in the years 1830 and 2000. Using these empirical case studies, we will analyse how changes in socioeconomic metabolism and land use are interrelated.

2.2   THE TRANSITION FROM AN AGRARIAN TO AN INDUSTRIAL SOCIOECOLOGICAL REGIME

The Transformation of the Energetic Metabolism

This section starts with an analysis of energy provision and explores the changes in the structure and quantity of the Austrian supply of primary energy during the last 170 years. This analysis will take into account not only the different types of primary energy carriers (that is, biomass, coal, oil, gas and water) but also the provenance of the respective primary energy, that is, the relation of domestic supply to imports and exports of energy. The discussion of energy provision is followed by a rough analysis of how the available primary energy is converted into final energy and how it is used to provide different energy services. Finally, we relate the changes in energy consumption to the growth of GDP.

Energy provision

The time series (Figure 2.2) shows the gradual transition of Austria from an agrarian, biomass-based energy system to an industrial energy system highly dependent on fossil fuels and, to some extent, on hydropower, which is abundant in this country due to its mountainous terrain and humid climate.

For the time series of energy flows this research drew mainly upon official Austrian statistics (energy, mining, agricultural and trade statistics), which provide comparable data from the early 19th century onwards

a) GDP, population

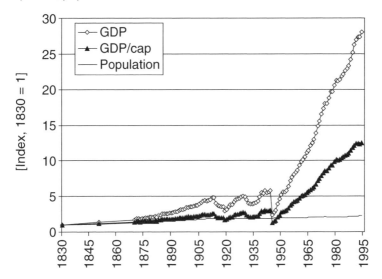

b) Population breakdown

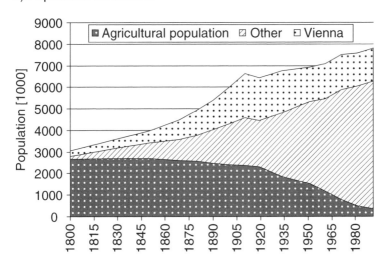

Notes:
a) GDP, GDP per capita, population (index, 1830 = 1).
b) Population breakdown: rural/Vienna/other; the latter category includes population living in all of Austria's cities and towns excluding its capital.

Figure 2.1   Basic indicators for Austria's development from 1830 to 2000

a) Total energy consumption (DEC)

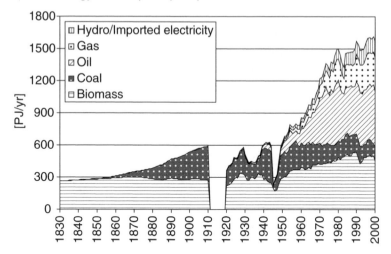

b) Per-capita energy consumption

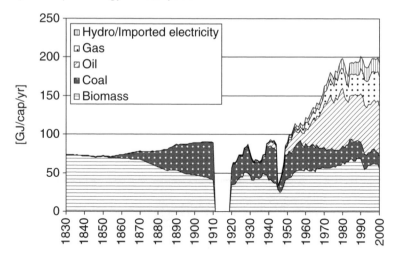

Notes:
a) Total energy consumption of Austria.
b) Per-capita energy consumption.

Source: Krausmann, Schandl and Schulz (2003).

Figure 2.2   Domestic energy consumption of Austria, 1830–2000,
including biomass for human food and animal nutrition, and so
on (see text for methodological explanation)

(see ÖSTAT 1979). Wherever statistical data were not available (for example, certain fractions of biomass extraction, the use of hydropower in the 19th century, feed for draught animals), and for cross-checking data, we used secondary sources and modelling approaches.[5]

Figure 2.2 shows Austria's domestic energy consumption (DEC) from 1830 to 2000, both as total energy flow and as per-capita figures. As usual, domestic consumption is defined as domestic extraction plus import minus export (that is, 'apparent consumption'). Austria's overall DEC increased by a factor of 6 from 1830 to 2000 and per-capita DEC increased by a factor of 2.7. This rise in energy consumption was made possible by the increasing use of fossil fuels. The mix of fossil fuels used was similar to the general pattern in most industrial countries: in a first phase, industrialization was powered by coal, which substituted for wood. The use of oil began gradually around 1900 and increased quickly after World War II. Natural gas – the large-scale use of which requires a widespread grid of pipelines – started to gain considerable importance after the 1970s.

Hydropower development had already begun in the early 20th century. At present, about three-quarters of Austria's techno-economically feasible hydropower potential are used, contributing some 65 to 70 per cent to Austria's electricity generation or about 13 per cent to Austria's technical primary energy consumption. Hydropower is more important for Austria's energy system than these figures suggest: producing the same amount of electricity in fossil-fuel-fired power plants – which are considerably less efficient than hydropower plants for thermodynamic reasons – would require two to three times more primary energy (coal, oil or gas) than the amount of hydropower used.

Total biomass consumption increased by a factor of 1.7 from 1830 to 2000, while per-capita biomass DEC decreased by about one-quarter. Whereas biomass consumption remained more or less constant (per-capita use declined significantly) until 1950, it rose again after World War II and thereafter almost doubled by the 1990s. Per-capita use of biomass did not quite reach the 1830 value again. As will be discussed in more detail below, this increase in socioeconomic biomass flows was achieved through fossil-fuel-based agricultural intensification.

Figure 2.2 has shown the transition from an area-based to a fossil-fuel-based energy system. Austria has only minor domestic deposits of coal, oil and gas. Coal resources are restricted to rather small deposits of low-quality coal (lignite). Oil and gas deposits are also only small and contribute today only about 10 to 20 per cent of domestic energy consumption. Therefore, the transition to a fossil-fuel-based energy regime also entailed a considerable reliance on imports.

To discuss this issue it is important to remember that the 'Austria' to which we refer is a constructed entity for the period before World War I, not a nation state. Although coal used in the 19th century was 'imported' into this territory, it actually came from other parts of the same country, that is, from the large coal-mining districts in Bohemia and southern Poland, which also belonged to the Austro-Hungarian Monarchy. After World War I, the access of the newly-formed Republic of Austria to these now-foreign coal-mining districts was much more difficult and expensive. This led to a restructuring of Austrian industry, entailing, above all, a reduction of the energy-intensive iron industry (Krausmann and Haberl 2002). This explains why Austrian coal consumption peaked before World War I.

Figure 2.3 accounts for physical imports and exports of energy and biomass from 1920 onwards. Unfortunately, no detailed import/export data exist before 1920. It shows that Austria was a net importer of energy, while the trade balance of biomass was near zero for the entire period. Austria's dependency on imports increased from almost nil in 1830 to 60 per cent of total energy input and to almost 90 per cent of technical energy input in the late 20th century. Biomass imports and exports – both of which increase exponentially – are of almost equal magnitude, so that Austria's domestic extraction of biomass is similar to its domestic consumption even though current imports and exports of biomass amount to about 250 peta-joules per year (PJ/yr), which is about half the value of the domestic consumption of biomass (Figure 2.3b).

Austria's energy system in the early 19th century clearly bears the features of an agrarian regime. In 1830, primary energy use per unit of land amounted to 30 GJ/ha and was well within the physiological limits of the agrarian regime. In contrast, current energy consumption amounts to approximately 200 GJ/ha, which is far beyond the limits of a biomass-based controlled solar energy system: the energy currently consumed in Austria roughly equals the annual aboveground NPP of a forest covering the whole Austrian territory, or the sustainable yield of a forest three times the size of Austrian territory. Not only energy use increased but material use multiplied as well. Total DMC grew by a factor of 7 and per-capita DMC increased threefold to currently more than 18 tons. Despite an overall growth of biomass use, its share of total DMC declined from more than 90 per cent to 25 per cent.

In the Austrian case we observe a typical transition pattern from a comparatively stable level of energy use in the period before 1850, to a period of exponential growth lasting until the 1970s (disregarding the effect of the collapse of the Austro-Hungarian Empire and the fluctuation caused by the two World Wars and the economic crisis of the 1930s, which were obviously compensated for by rapid growth after World War II). The period of

a) Energy total

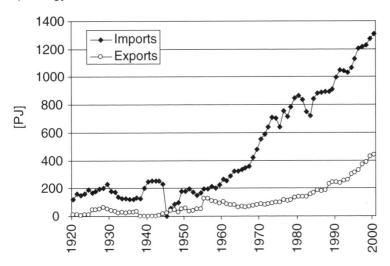

b) Biomass

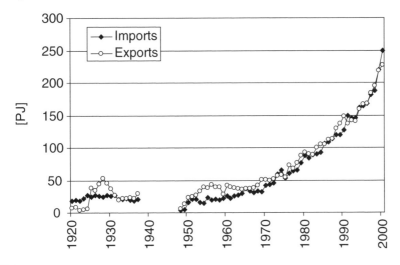

Notes:
a) Total primary energy.
b) Biomass.

Sources: Erb (2004); Krausmann, Schandl and Schulz (2003).

Figure 2.3 Austria's energy imports and exports, 1920–2000

growth that accompanied the transformation of the energy system was redeemed by a relative stabilization of per-capita energy use induced by the oil crisis in the 1970s.[6]

This transition process was fuelled by cheap and abundant fossil fuels. Our results support the assumption that the transformation of the energy system is a fundamental characteristic of the transformation of the socio-ecological regime and the whole process of industrialization, as it abolishes the traditional limitations for physical growth in agrarian societies: in contrast to biomass, fossil fuels are point sources with high energy density,[7] their exploitation is not limited by ecosystem properties (renewing rates, soil fertility), they fuel their own transportation and they open up a whole range of new energy technologies (for instance, the internal combustion engine). Among the major components of this transition process are: 1) unprecedented growth in energy demand; 2) introduction of fossil fuels and 3) regional displacement of supply and demand (that is, the globalization of the energy system).

Energy use
This section explores the transformation of the energy system with respect to the conversion of primary energy into final and useful energy and the provision of energy services. This allows more detailed insights into structural changes occurring with the implementation of the fossil energy system. Particularly with respect to socioeconomic growth, the efficiency and patterns of energy use are of paramount importance. The data on final and useful energy, however, are not available in annual resolution for the whole period but only for two points in time, 1830 and 1995. Figure 2.4 presents a tentative analysis of final energy use and useful energy produced in Austria in 1830 and 1995. Data were taken from Krausmann and Haberl (2002), where methodological aspects and data sources are described in detail. Although considerable uncertainties exist in these calculations, the fundamental patterns can be assumed to be reliable.

We find that the total per-capita use of final energy rose by a factor of 2.7, which is equal to the increase in per-capita DEC. This is a counterintuitive finding, because the efficiency of most energy conversion processes has risen considerably during this 165-year period.[8] That the aggregate conversion efficiency remained stable despite these technological improvements can partly be explained by two significant factors. 1) The aggregate conversion efficiency of an important conversion process, namely food production, has substantially declined. Animal products account for a larger part of the human diet in 1995 than in 1830, and much energy (about 90 per cent; Leach 1976) is being lost in the conversion from plant biomass to meat, milk and eggs. 2) The proportion of final energy constituted by

a) Final energy used

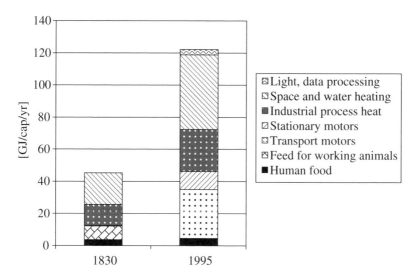

b) Useful energy delivered

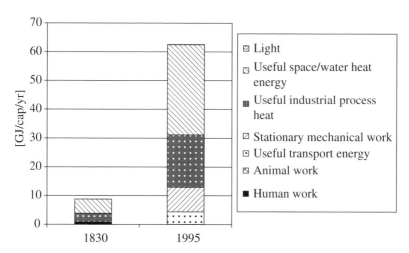

Source: Krausmann and Haberl (2002).

Figure 2.4   Useful energy analysis of Austria, 1830 and 1995

processed forms of energy, above all, electricity, increases. Before 1900, the only conversion process from primary to final energy was food production and minor amounts of charcoal production, while all other kinds of energy were used in the form in which they were extracted. As the share of processed energy rises, losses during conversion from primary to final energy are also bound to increase, even if the efficiency of each conversion process (for example, electricity generation) improves.[9]

In 1830, more than 70 per cent of the final energy was used for process heat and space heating, the remainder (27 per cent) being used mostly for nutrition for humans and working animals, that is, the final energy available to deliver drive power. According to our estimates, wind and water mills, while delivering decisive services, probably accounted for less than 1 per cent of the final energy use. Use of wind for transport purposes (sailing boats) was not accounted for but was probably of minor importance for this land-locked territory. Use of hydropower for transport on rivers was probably more important but this too could not be quantified. In 1995, by contrast, nutritive energy accounted for less than 4 per cent of the total. The most important use of final energy was space and water heating (38 per cent), followed by transport (25 per cent) and industrial process heat (21 per cent).

If we look at the useful energy delivered, efficiency gains become clearly visible. The total amount of useful energy available per capita was seven times greater in 1995 than in 1830. Here increases in the efficiency of conversion processes (such as more efficient motors, furnaces or heating systems) play a decisive role. Moreover, the substitution of internal combustion machines or electric motors for working animals and human labour greatly increased the efficiency of drive-power generation and transport: while draught animals may reach maximum conversion efficiencies of 20 to 30 per cent while performing work, their overall (lifetime) efficiency is much smaller and usually does not exceed a few per cent. In contrast to machines, draught animals have to be fed for quite some time before they can be used to perform work and they can not be 'switched off' during periods in which they do not work. In 1830, food and feed, the final energy to be converted into animate power, constituted more than a quarter of total final energy supply, while work of humans and animals accounted for a significant 7 per cent of delivered useful energy. In terms of useful energy, however, water power also gains significance. The contribution of water-driven stationary motors amounted to almost a fifth of the useful energy delivered by humans and animals together.

Both human and animal power lost their significance during the transformation of the energy system. While human-generated physical work became almost immeasurably small and animal power vanished completely, drive power generated by stationary motors (mostly electric motors) was much more important than their share in final energy consumption (14 per cent instead of 9 per cent). This reflects the high quality of electricity as a final energy carrier: while internal combustion machines seldom reach conversion efficiencies above 35 per cent, due to the second law of thermodynamics, electric motors may even convert as much as 95 per cent of the electricity into drive power. By far the largest share of useful energy was, similar to 1830, used for space and water heating (50 per cent), followed by industrial process heat (29 per cent).

This comparison between 1830 and 1995 shows that the significant increases in the conversion efficiency from primary into final and useful energy did not lead to any reduction of primary energy input but to an enormous growth of useful work and available energy services. Assuming that animal work had been mainly used as draught power – that is, for transportation purposes – in 1830, we may derive the growth factors shown in Table 2.1.

Table 2.1 shows that by far the largest increase (namely 70-fold) was that of stationary motors, that is, devices to drive machinery. Transportation energy grew almost 15-fold. In both cases, machine power was substituted for human and animal work, the former of which decreased to one-tenth and the latter to (almost) nil. Growth in the amount of useful energy delivered per capita as process heat and for space/water heating was approximately six fold. These data point to the outstanding significance of transport and machine-powered work for the transformation of the socioecological regime. At the same time, the sixfold increase in availability of heat for industry and households is remarkable and indicates the structural differences in social metabolism associated with the transformation of the energy system during the transition from an agrarian to an industrial regime.

It would probably be rewarding to look into the time path of these changes but at present there are unfortunately no sufficiently reliable data

Table 2.1   Growth in the per-capita availability of useful energy in Austria from 1830 to 1995

	Growth Factor
Human work	0.1
Transportation energy	14.9
Stationary motors	71.3
Industrial process heat	6.0
Space and water heating	6.4
Light	n.d.

Source: Krausmann and Haberl (2002).

to do so. For example, one could probe whether the reduction in the amount of physical human work was a smooth, continuous process or whether human labour energy expanded in the first phase of industrialization and fell only afterwards. It has been argued, for example, that the Industrial Revolution was indeed an 'Industrious Revolution' (DeVries 1994), implying more (and maybe even harder) working hours for the population. Empirical evidence for the Austrian case with respect to the size of labour force and number of working hours indeed supports this assumption. These data indicate that during the coal and steam engine period of industrialization, the demand for labour increased with coal consumption and industrial output. Only the second phase of the transformation process, based on oil and electricity, resulted in an absolute substitution of human work by fossil fuels and a significant decline in the demand for human labour in production.[10]

Energy and GDP
Real per-capita GDP grew by a factor of 12.2 (Haberl and Krausmann 2001) between 1830 and 1995. If we compare the growth in GDP and in energy use, we find that GDP grew considerably faster than any indicator of aggregate energy throughput, even useful energy. The growth in useful energy was only about half that of GDP growth (about the same applied to process heat and space/water heating). By contrast, useful energy for stationary motors grew about six times faster than GDP. Useful energy for transport grew at a similar rate to that of GDP. This implies that the amount of useful energy required per ton-kilometre must have been reduced considerably as freight transport seems to have increased by two orders of magnitude during transitions from agrarian to industrial society (Fischer-Kowalski, Krausmann and Smetschka 2004). Less than one-hundredth of the amount of physical human work required in 1830 per unit of GDP was needed in 1995, thus impressively demonstrating the extent to which machine power was substituted for human labour.

The increases in energy conversion efficiency were more than compensated by GDP growth. That is, at least over the time period observed, the introduction of more energy-efficient technology served rather to fuel GDP growth than to reduce aggregate energy consumption. This pattern might change, however, with a growing importance of (rather energy-extensive) information processing services for growth (Ayres and van den Bergh 2005). Significant changes in energy prices, either due to resource scarcity ('peak oil') or due to socioecological tax reforms that increase tax burdens on energy use while reducing taxes on salaries, might also reverse that trend.

Changes in Land Use During the Transition

Land use is the most important human activity with respect to energy provision in the agrarian regime. Energy availability in a controlled solar energy system is determined by a number of closely interlinked factors: 1) the amount of land available per capita;[11] 2) the productivity of the land;[12] 3) the spatial pattern of energy demand and 4) the availability of biomass.[13] Much of what we know about land use in Austria in 1830 suggests that these were the major bottlenecks of growth at that time and were reflected in land use and landcover patterns.

In Austria, cropland agriculture prevailed in mountainous regions even at high altitudes (above 1000 m a.s.l.) under unfavourable conditions because the staples were needed to nourish the local population. Even in the most fertile lowlands on the other hand, livestock was kept not only to provide working animals but also to collect scarce nutrients from extensively used grassland and distribute them on the best, respectively most intensively used, plots of land. The agricultural literature, cadastral data and the first forest inventories available (see Chapter 3) suggest that forests were used intensively, maybe even excessively, not only to harvest wood but also for feed and litter, thus removing considerable amounts of nutrients from forests. According to our data, the early stages of industrialization did little to change this state of affairs, whereas later stages, above all after 1950, completely altered the picture. This is consistent with many indicators that show that the industrialization of Austria's agriculture that had started slowly between the World Wars was halted by World War II but was then completed within a few (two to three) decades afterwards (Krausmann 2001).

Cropland area seems to have reached a peak in the late 19th century, consistently and considerably decreasing thereafter (Table 2.2). From 1910 to 1995, cropland areas decreased by more than one-third and grassland and alpine pastures by about 10 per cent. Austria's total agricultural area fell by over 9000 km^2 between 1830 and 1995 – more than 10 per cent of Austria's total area. Roughly one-third of this decrease was consumed by the growing area demand of settlements, infrastructure and production sites and roughly two-thirds converted to forests. Alpine pastures and natural alpine vegetation remained about constant.

One important partial explanation for these patterns is that changes in agricultural practices and technology, above all, fertilization, irrigation and the use of improved crop varieties, resulted in considerably increasing yields. Commercial yields of the most important crops increased by factors of about 3–6 (see Figure 2.5), and by and large, food output kept pace with population growth during the 19th century and even exceeded demand after World War II.

Table 2.2   Changes in landcover, Austria 1830–1995

	Built-up Land[a] [km²]	Cropland [km²]	Grassland [km²]	Alpine Pastures [km²]	Forests [km²]	Alpine Vegetation [km²]
1830	696	17 669	15 441	8 717	30 727	5 874
1880	889	18 798	12 588	9 708	31 820	5 323
1910	1 103	18 121	13 174	9 735	31 689	5 300
1950	1 594	13 522	16 044	8 731	33 343	5 896
1980	2 783	13 361	13 240	8 128	35 338	6 232
1995	3 261	11 591	12 291	8 520	37 362	6 060
Δ1995/1830 [km²]	+2 565	−6 078	−3 150	−197	+6 635	+186
Δ1995/1830 [%]	+368	−34	−20	−2	+22	+3

Note: [a] 'Built-up' land includes sealed soils and other urban areas, for example, parks, gardens, and so on.

Source: Krausmann (2001).

A host of technological changes contributed to this increase in yields. In the 19th century, the most important innovations were the introduction of new crops, above all, legumes and root crops, and the reduction of fallow from 15 per cent to about 3 per cent by 1910. The widespread cultivation of the potato, which yielded almost twice as much nutritional energy per hectare as rye, was also a major factor. Legumes were important not only as feed crops, but also because of their ability to fix nitrogen through symbiotic bacteria. Increased availability of feed crops enabled farmers to keep more livestock in stables, thus improving fertilization. This, together with the availability of more straw to be used as litter, considerably reduced pressures on forest ecosystems. Tentative calculations (Krausmann 2006b) indicate that nitrogen fertilization through legumes and dung on Austria's agricultural area increased by about 60 per cent from 1830 to 1910. It is important to note that 19th-century improvements in agricultural productivity by and large remained within the controlled solar energy system. Except for the increasing use of iron, agriculture was not subsidized by fossil energy. This is reflected in an observed increase in the energy efficiency of agriculture by a factor of 2 between 1830 and 1910 (Krausmann 2006b).

In the first half of the 20th century, the industrialization of agriculture proceeded slowly, if at all. Agricultural yields in 1950 were similar to those in 1910. Even in 1950, there were about 600 000 draught animals and only about 30 000 tractors. This picture changed completely after 1950: within

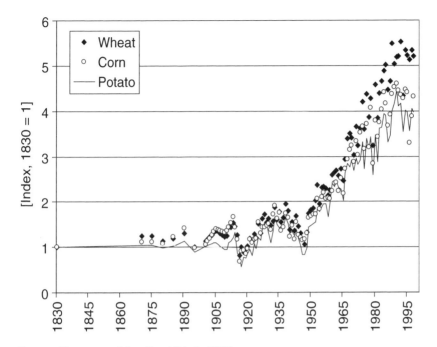

Source: Krausmann, Schandl and Schulz (2003).

Figure 2.5   *Yields of important crops in Austria, 1830–1995 (index, 1830 = 1)*

only two decades, from 1950 to 1970, draught animals disappeared almost completely, the number of tractors increased tenfold and the application of nitrogen in mineral fertilizer by a factor of about 8. In the same period, the number of agricultural workers decreased by about 60 per cent.

Table 2.3 summarizes changes in Austria's agriculture for the whole period. It shows that the number of agricultural workers decreased steadily, but at different rates: from 1830 to 1910, the agricultural workforce fell slowly (about −16 per cent in 80 years), while it fell by 20 per cent in only 40 years from 1910 to 1950, by 60 per cent in only 20 years from 1950 to 1970 and by another 65 per cent in 30 years from 1970 to 2000. Across the whole period, the number of agricultural workers fell by 92 per cent. Draught animals peaked in 1910 and were about as numerous in 1950 as they had been in 1830. After 1950, they all but vanished in only 20 years. Installed power – that is, the aggregate maximum work capacity of all agricultural workers, working animals and machinery – remained about constant for 120 years from 1830 to 1950, surged by one order of magnitude from 1950 to 1970 and more than tripled afterwards until the year 2000.

Table 2.3   Indicators for changes in Austria's agriculture from 1830 to 2000

	Unit	1830	1870	1910	1950	1970	2000
Agric. labour force	[1000]	1650	1520	1370	1092	432	150
Draught animals	[1000]	554	661	733	580	25	0
Tractors	[1000]	0	0	0	30	268	336
Installed power[a]	[GW]	0.5	n.d.	0.6	0.5	5.6	17.1
Mineral fertilizer	[kt N/yr]	0	0	0	54	438	230
Grain yield	[t/ha]	0.9	1.1	1.2	1.6	3.2	5.5
Area productivity	[GJ/ha][b]	3.0	3.7	5.3	5.7	11.3	16.1
Labour productivity	[GJ/cap][c]	8	10	15	22	100	354

Notes:
[a]   Installed power is defined as the power of draught animals (0.7 kW per horse, 0.5 kW per oxen), agricultural workers (0.1 kW per person) and agricultural machinery, above all, tractors.
[b]   GJ nutritional value (which is a bit lower than the calorific value) of final products of the agricultural sector per hectare of agricultural area (i.e., cropland, managed grassland).
[c]   GJ nutritional value per agricultural worker.

Sources: Krausmann (2001, 2006b).

Grain yields increased by some 80 per cent from 1830 to 1950 but by a soaring 244 per cent in only a half-century afterwards. The amount of nutritional energy produced per hectare of agricultural area (that is, cropland plus managed grassland) less than doubled in 120 years from 1830 to 1950 but almost tripled in the 50 years that followed. Labour productivity, defined as amount of nutritional energy produced per agricultural worker, grew by a factor of less than 3 from 1830 to 1950, followed by a staggering 16-fold increase from 1950 to 2000.

These data support Grübler's (1998) view that agriculture improves in a first phase primarily through biological innovations such as new crops and new agricultural practices. The second phase distinguished by Grübler – the expansion of agriculture on continental and even global scales through new transport technologies – cannot be seen in our example, because Austria's total area was already under cultivation in 1830 and throughout the 19th century. In this period, Austria hardly participated in the increasing global food trade compared with other European economies. Urbanization – which also depends on innovations in the transport system – is, however, also proceeding in Austria as shown in Figure 2.1. The third phase, characterized by mechanization, use of synthetic fertilizers, pesticides and use of new crops, proceeded quickly and with far-reaching consequences in Austria. The fourth and final phase distinguished by Grübler (1998), reduction of the amount of cultivated land achieved through a combination of saturation of demand, increasing international trade and further yield increases, is also clearly visible in Austria (see Krausmann and Haberl et al. 2003).

These technological changes had many important implications (see, for example, Grübler 1998). We will focus here on changes they brought about in the socioecological regimes. It must be remembered that in 1830, Austria's energetic metabolism almost entirely depended on harnessing the productivity of plants growing on the country's own territory, in other words, on its NPP. In fact, almost all of Austria's area suitable for cultivation (except for very elevated areas, inaccessible terrain or wetlands) had been converted from natural ecosystems (mostly forests) into managed or 'colonized' ones, to cropland, pastures and grazing land and managed forests. That these colonized ecosystems produced more of those kinds of biomass useful for humans, such as food, feed and fibres, does not imply that they became more productive in a biological sense. On the contrary: the agricultural practices in the 19th century and the concomitant changes in landcover resulted in a considerable reduction of the productivity of ecosystems.

This can be seen by looking at changes in the 'human appropriation of NPP' or HANPP (Haberl 1997; Vitousek et al. 1986; Wright 1990).[14] The aboveground NPP of the potential vegetation of the Austrian territory amounts to about 1400 PJ/yr or 150 million tons of biomass. This exceeds the amount of energy used in 1830 about fivefold and roughly corresponds to Austria's current domestic energy consumption (see Figure 2.2).[15] As already indicated above, biomass harvest rose considerably, from 294 PJ/yr in 1830 to 507 PJ/yr in 1995, that is, by +72 per cent. In other words, it increased from 21 per cent to 36 per cent of the potential NPP. Counter-intuitively, this surge in biomass harvest did not result in a devastation of Austria's ecosystems because the technological changes discussed above resulted in a considerable increase not only of commercial yields per unit area but also of the productivity of Austria's vegetation: the aboveground NPP on Austria's territory increased from 859 PJ/yr in 1830 to 1201 PJ/yr in 1995, that is, by +40 per cent (Table 2.3). In absolute figures the increase in NPP (+342 PJ/yr) was even larger than the increase in biomass harvest (+213 PJ/yr). This explains why the total percentage of NPP appropriated in Austria fell from 60 per cent to about 50 per cent of potential vegetation, even though biomass harvests rose substantially, both in absolute figures and as a percentage of actual NPP.

From a broader perspective we can therefore see that a transition from an agrarian to an industrial energy system results in a considerable degree of 'decoupling' between economic growth and area demand for agriculture.

Table 2.4   Aboveground NPP, harvest and HANPP in Austria, 1830–1995

	NPP_{act} [PJ/yr]	NPP_{act} [% of NPP_0]	Harvest [PJ/yr]	Harvest [% of NPP_{act}]	HANPP [PJ/yr]	HANPP [% of $ANPP_0$]
1830	859	61	294	34	839	60
1880	911	65	305	34	798	57
1910	948	68	310	33	765	55
1950	991	70	296	30	708	50
1980	1198	85	484	40	690	49
1995	1201	86	507	42	710	51
Δ1995/1830 [PJ/yr]	+342		+213		−129	
Δ1995/1830 [%]	+40		+72		−15	

Source: Krausmann (2001).

This decoupling takes place not only because fossil fuels replace biomass but also because fossil-fuel-powered technology allows a surge in area efficiency. A good indicator of area efficiency is the ratio of harvest to HANPP. For example, in Austria in 1830 this ratio was 0.35, implying that harvest of biomass for human purposes accounted for only one-third of total HANPP, whereas two-thirds were caused by the prevention of NPP, that is, by human-induced reductions in the productivity of vegetation. In 1995, this ratio had climbed to 0.71 – harvest accounted for more than two-thirds of HANPP and productivity losses for less than one-third.

It may be interesting to note that settlements, infrastructure and other sealing of soils contributed less than 2 per cent to total HANPP in 1830 but climbed to about 6 per cent by 1959 and to about 8 per cent by 1995 (Schulz, Krausmann and Haberl 2001). That is, by 1995 soil sealing contributed to the prevention of NPP already by about one-third and agricultural land use by two-thirds. Contemporary agro-ecosystems are sometimes even more productive than the natural vegetation they replace in terms of NPP in some intensively cropped regions, above all, those dominated by C4 plants such as maize. In such regions area efficiency, defined as above, may be greater than 1, that is, harvests may be larger than aboveground potential (natural) vegetation.[16]

To summarize, the figures presented here confirm the hypothesis outlined in the beginning of this chapter that industrialization fundamentally changes the role of land use: land availability ceases to be a factor constraining economic or even population growth. Agriculture ceases to be the decisive source of socioeconomic energy: in terms of final products, it may even deliver less or at least not much more energy than it consumes.

The availability of area-independent fuels allows a substantial energetic 'subsidy' for the agricultural system, thus boosting yields and even the net primary productivity of the vegetation. This allows forests to re-grow, even if urbanization and other infrastructure development claim increasing amounts of area, resulting in a surge of area efficiency that even leads to a downward overall trend of HANPP.

Mechanization and other technological improvements allowed for a staggering 44-fold increase in labour productivity, that is, the amount of nutritional (food) energy produced per agricultural worker. One aspect of the price to be paid for these improvements is the deterioration of the energetic return on investment in agriculture. Others include the increased use of agrochemicals, growing field sizes and large monocultures, resulting in deterioration of the ecological and aesthetic value of agricultural landscapes (Peterseil et al. 2004; Wrbka et al. 2004). Yet another of the many impacts this change produced will become clear in Chapter 5, when we discuss the relations between different scales, that is, the relations between rural periphery and urban centres, and the changes in biomass (and nutrient) flows they entail.

2.3   CONCLUSIONS

Looking at the development of energetic metabolism in Austria since the early 19th century, industrialization appears as a stepwise process of de-linking energy provision from land. In a first phase, beginning in the 1840s and lasting until World War II, coal was substituted for biomass with respect to the provision of heat and, with the gradual improvement of the steam engine, also for work and motion and allowed for an overall growth in energy use. This process, however, did not result in a marginalization of animate power. Coal-powered steam engines only brought about a relative substitution of fossil fuels for animate power but the absolute demand for labour increased. Throughout the 19th century, agriculture did not receive substantial subsidies from the fossil fuel energy system. The provision of food and feed basically remained within the limitations of the controlled solar energy system. Only the second phase of this transition process, which was fuelled by oil and later by natural gas and closely related to technological possibilities provided by electricity and the internal combustion engine, brought about a complete implementation of the fossil fuel energy system and a de-linking of socioeconomic metabolism from land. While the first phase of the transformation process was by and large restricted to coal-consuming urban/industrial centres interlinked by a railroad network but embedded in an agrarian matrix, during the second phase of the transition all sectors of the economy and everyday life were fully penetrated by the fossil fuel energy system. This triggered growth of energy and material consumption not only in the industrial sector and transport but also in households to a level far beyond what would have been possible under the conditions of the agrarian regime.

The transition from the agrarian to the industrial socioecological regime implies changes in resources used, in technology, in settlement patterns, in urban–rural relations and in land use. One fundamental change refers to the main resources used: during industrialization, fossil fuels are increasingly used and biomass loses its role as the main energy source of society. Biomass extraction depends on ecological primary production (net primary production, NPP), that is, the synthesis of organic matter (biomass) from inorganic substances (mainly CO_2 and H_2O) through photosynthesis. NPP is an inherently area-dependent process that converts radiant energy from the sun into energy stored chemically in biomass. While the agrarian energy system relies almost exclusively on harnessing this process for human needs, industrial metabolism depletes concentrated stocks of resources that had earlier been accumulated by geological or combined biogeological processes. The extraction rate of the latter resources is not constrained by land availability. The transition to an industrial energy system thus removes tight limits on growth imposed on agrarian societies by their dependence on the productivity of local ecosystems. The transition from biomass to fossil fuel energy (and later nuclear energy, large-scale hydropower and new renewable sources) was a precondition for the rapid economic (monetary and biophysical) growth in Europe from the 18th century onwards (Pomeranz 2000; Sieferle 1997, 2001; Smil 1992).

The significance of land use for socioeconomic metabolism changes qualitatively and quantitatively during industrialization. For an agrarian society, agriculture is the central source of energy. The agricultural sector as a whole must therefore maintain a positive energy balance, that is, it must supply more energy to society than society invests in it. Under agrarian conditions, the energy return on investment (EROI) (Hall, Cleveland and Kaufmann 1986) must be – and has thus far been consistently found to be – between 5 and 10, otherwise an agrarian society cannot survive.[17] Industrial societies, by contrast, can afford to 'subsidize' their agricultural sector energetically and very often do so. In Austria, for example, the aggregate EROI of agriculture was about 1 (implying that as much energy was invested as gained) for the whole period from 1950 to 2000 (Krausmann and Haberl et al. 2003). Of course, many final products such as meat, milk and so on, have EROIs of considerably less than 1.

The removal of the tight limits to energy availability that were characteristic for agrarian societies was a prerequisite for the development of new agricultural technologies that allowed substantial increases in yields, that is, in the amount of agricultural products harvested per unit of area (Grübler 1998; Hall, Perez and Leclerc 2000). It was also a prerequisite for spectacular improvements of labour efficiency, that is, the amount of food energy produced per agricultural worker, by more than two orders of magnitude. While the improvements of yields were a prerequisite for the growth in population and in per-capita consumption during industrialization, the improvement of labour productivity was a necessity to satisfy the growing demand of the manufacturing and services sectors for human labour power.

Fossil and nuclear fuels as well as large-scale hydropower are highly concentrated and can be extracted with a high EROI (Hall et al. 1986). Fossil fuels can be conveniently stored, easily transported and converted into useful energy with comparatively high efficiency. This allowed the development of new transport systems with far-reaching consequences for the spatial organization of societies. Among others, it was a prerequisite for current patterns of urbanization, spatial organization of production and consumption, the spatial separation of different agricultural production processes such as livestock rearing and cropland farming, and many other environmentally highly relevant phenomena. Growth in transport volumes by far exceeds the growth in energy or materials throughput and a considerable fraction of this growth is the inevitable consequence of urbanization.

For the Austrian case, the changes in social metabolism related to industrialization show a typical transition pattern. In the early 19th century, per-capita energy throughput started to grow and the steady-state agrarian regime slowly began to change. The following period of roughly 130 years is characterized by a transition process towards the industrial socioecological regime, during which the old physical boundaries were gradually overcome. This process was related to unprecedented growth of the physical economy powered by fossil fuels, first in the form of coal and later on as petroleum. Material and energy turnover, both in absolute terms and per capita, multiplied during this period. By the 1970s, the penetration of the industrial regime was complete, all limitations of the agrarian regime had disappeared and growth slowed down. Thereafter, material and energy use per capita seem to have stabilized at a typical industrial level. Absolute growth, however, continues, although at a significantly slower pace than in the decades before. This stability is, however, of a different quality than the stability observed for the agrarian regime: in contrast to the agrarian regime, which is based on renewable energy, the current regime relies upon a high throughput of a finite energy source, the day upon which this source is exhausted being already in sight. But not only the source but also the sink is finite: the capacity of ecosystems to absorb the enormous waste and emissions output of industrial societies is increasingly exploited to the limits.

From this perspective, the industrial regime appears to be rather an intermediate period in a still-ongoing transition process than a stable endpoint.

ACKNOWLEDGEMENTS

This chapter draws on material from a number of empirical studies on the long-term historical relation of land use and social metabolism in Austria. Parts of the research behind this chapter were funded under the research programme 'Austrian Landscape Research' (http://www.klf.at/) of the Austrian Ministry of Education, Science and Culture and the German Breuninger Foundation's programme 'Europe's Special Course into Industrialization' in cooperation with Helga Breuninger and Rolf Peter Sieferle. Finally, most of the analytical work and the synthesis presented in this chapter was undertaken within the Austrian Science Fund project 'The Transformation of Society's Natural Relations' (Project No. P16759).

Methodological aspects and detailed analysis of the data used in this chapter have been published in a number of books and journal articles, above all, Krausmann (2001, 2004); Krausmann and Haberl (2002); Krausmann and Haberl et al. (2003); Krausmann, Schandl and Schulz (2003); Krausmann et al. (2004) and the book by Sieferle et al. (2006).

The authors are grateful to Rolf Peter Sieferle, Marina Fischer-Kowalski and Heinz Schandl for their valuable contributions and frequent discussions vital to the progress of our work, as well as to Helga Weisz, Karl-Heinz Erb and Niels Schulz for their useful comments at various stages of the project.

NOTES

1. All data we report for the time period before 1918 refer to a territory of 85 906 km², of which almost 95 per cent are within Austria's current borders. Six thousand km² of this territory today form part of Italy and Slovenia. Furthermore, the modern Austrian province of Burgenland is not considered in pre-1910 analysis because it was not possible to get data for this province for the 19th century. Burgenland has a total area of 3900 km² (4.7 per cent of the current Austrian territory) and was under Hungarian administration until 1918. All data for the period 1918–95 include Burgenland and refer to today's Austrian territory of about 83 400 km². The creation of such a consistent database was made possible by the Austrian Science Fund (Project No. P16759-G04), with the aim of reconstructing the relevant biophysical, economic and social data for the Austro-Hungarian Empire as a whole. This is still ongoing.

2. Of course, agrarian regimes too can experience periods of physical growth and growth processes have occurred historically in agrarian societies (Sieferle 2001). Growth under the conditions of the agrarian regime is bound to 1) increasing the amount of land under cultivation/management and/or 2) optimizing the efficiency of the land-use system (for example, by shifting from a long fallow system to short fallow or crop rotation (see, for instance, Boserup 1965). Both preconditions are, however, limited by physical

constraints, and growth in agrarian regimes tends to be a process of approximation to the elastic border of a steady state.

3. In economic history, the term 'mineral economy' (Wrigley 1988) has gained popularity.

4. Energy metabolism, as an element of the concept of social metabolism, is defined here in a most comprehensive way (Haberl 2001). While conventional measures of energy consumption (IEA 2004; UN 2002), with their focus on technical energy use, allow one to differentiate well between industrial societies, they do not adequately capture the energetic base of pre-industrial societies as they ignore human and livestock nutrition. In contrast, the concept of energy metabolism used here is broad enough to 1) be applied to all types of socioecological regimes and 2) allow socioeconomic energy flows to be related to energy flows in ecosystems. These energy flow accounts are methodologically compatible with material flow accounts (MFA) in their most current versions (Eurostat 2001). This means, above all, that all biomass flows are accounted for, regardless of whether the biomass is used in technical installations such as furnaces or biogas plants or for human nutrition, as animal feed or for other purposes. In analogy to MFA, EFA (energy flow accounting) allows the assessment of the following indicators: 'direct energy input' (DEI) is analogous to the direct material input (DMI), an indicator for the sum of energy (or material) inputs of a defined socioeconomic system. DEI is defined as the sum of domestic energy extraction plus energy imports. DEI minus energy contained in exports is referred to as domestic energy consumption (DEC), the corresponding indicator to domestic material consumption (DMC). This close methodological and conceptual interlinkage between material and energy flow accounting is what we refer to as 'MEFA' (material and energy flow accounting) methodology (Krausmann et al. 2004).

5. A detailed description of data sources and modelling techniques is provided in Krausmann (2006a); Krausmann and Haberl (2002); Krausmann and Haberl et al. (2003).

6. Energy consumption in absolute terms, however, continued to grow during this period, but at a significantly slower pace (see Figure 2.2).

7. The energy density of biomass ranges between 5 and 15 megajoules per kilogram (MJ/kg), the energy density of lignite is between 10 and 20 MJ/kg and that of high-quality fossil fuels between 30 (hard coal) and 45 (petroleum, natural gas) MJ/kg.

8. For example, the conversion efficiency of electricity production in thermal power plants increased from less than 10 per cent in the 1930s to more than 40 per cent in the 1970s.

9. Moreover, no significant number of working animals existed anymore in 1995. According to the accounting method used (Haberl 2001), feed for working animals is counted as 'final energy', whereas feed for all other livestock is regarded as an internal flow of the agricultural sector needed to produce food for humans (and therefore not accounted for as 'final energy'). As a result, the share of plant materials in final biomass consumption drops as working animals vanish. A fourth factor is the increasing share of primary energy used for non-energetic purposes. These amounts are included in the figures of primary energy but are not accounted for in the figures of final energy.

10. Based on data for labour force in agriculture, mining and industry, average working hours and power delivery (Bolognese-Leuchtenmüller 1978; Hwaletz 2001; Maddison 2001).

11. In the 19th century, population density in Central Europe was already high and practically all land was cultivated. However, the cultivation of vast areas in the 'New World', Russia and colonial territories in the second half of the 19th century was an important factor in providing expanding European economies with food, feed and raw materials (Pomeranz 2000; see also Chapter 5).

12. The productivity of the land was closely linked to population density and determined by climatic conditions, the type of land-use system and the technologies applied (see Boserup 1965).

13. Under the agrarian regime, the aggregate availability of biomass is of less importance than its regional distribution: due to the high energy costs of transport, it was crucial that the location of energy resources coincided with demand, or that water transport was available. This was of special importance for energy-demanding industries and urban centres but had a considerable impact on land-use patterns in general. Therefore, staple foods tended to be locally grown even under unfavourable climatic conditions.

14. Human appropriation of NPP (HANPP) is a prominent measure for the human domination of terrestrial ecosystems. It measures the aggregate effect of land-use changes and biomass harvest on biomass availability in ecosystems, or, in other words, the human impact on energy flows in ecosystems.

15. As data on belowground NPP are highly uncertain, our analysis was restricted to the aboveground compartment. This chapter reports data from a published study (see Krausmann 2001 for methodological details).

16. The inclusion of belowground processes changes the picture somewhat, as the belowground productivity of forests is by far larger than that of cropland.

17. See the research tradition of Pimentel (Giampietro 1997; Giampietro and Pimentel 1991; Giampietro, Bukkens and Pimentel 1997; Giampietro, Cerretelli and Pimentel 1992; Pimentel, Dazhong and Giampietro 1990; Pimentel et al. 1973).

REFERENCES

Ayres, Robert U. and Jeroen C.J.M. van den Bergh (2005), 'A Theory of Economic Growth with Material/Energy Resources and Dematerialization: Interaction of Three Growth Mechanisms', *Ecological Economics*, **55**, 96–118.

Bolognese-Leuchtenmüller, Birgit (1978), *Bevölkerungsentwicklung und Berufsstruktur, Gesundheits- und Fürsorgewesen in Österreich 1750–1850*, Wien: Verlag für Geschichte und Politik.

Boserup, Ester (1965), *The Conditions of Agricultural Growth. The Economics of Agrarian Change under Population Pressure*, Chicago: Aldine/Earthscan.

Boyden, Stephen V. (1992), *Biohistory: The Interplay Between Human Society and the Biosphere – Past and Present*, Paris, Casterton Hall, Park Ridge, New Jersey: UNESCO and Parthenon Publishing Group.

DeVries, Jan (1994), 'The Industrial Revolution and the Industrious Revolution', *Journal of Economic History*, **54**(2), 249–70.

Erb, Karl-Heinz (2004), 'Actual Land Demand of Austria 1926–2000: A Variation on Ecological Footprint Assessments', *Land Use Policy*, **21**(3), 247–59.

Eurostat (2001), *Economy-wide Material Flow Accounts and Derived Indicators. A Methodological Guide*, Luxembourg: Eurostat, European Commission, Office for Official Publications of the European Communities.

Fischer-Kowalski, Marina (2003), 'Ecology, Social', in Shepard Krech III et al. (eds), *Encyclopedia of World Environmental History*, London and New York: Routledge, pp. 396–400.

Fischer-Kowalski, Marina and Helmut Haberl (1997), 'Tons, Joules and Money: Modes of Production and their Sustainability Problems', *Society and Natural Resources*, **10**(1), 61–85.

Fischer-Kowalski, Marina, Fridolin Krausmann and Barbara Smetschka (2004), 'Modelling Scenarios of Transport Across History from a Socio-metabolic Perspective', *Review. Fernand Braudel Center*, **27**(4), 307–42.

Gellner, Ernest (1989), *Plough, Sword and Book. The Structure of Human History*, Chicago: University of Chicago Press.

Giampietro, Mario (1997), 'Linking Technology, Natural Resources, and the Socioeconomic Structure of Human Society: A Theoretical Model', *Advances in Human Ecology*, **6**, 75–130.

Giampietro, Mario and David Pimentel (1991), 'Energy Efficiency: Assessing the Interaction between Humans and their Environment', *Ecological Economics*, **4**(2), 117–44.

Giampietro, Mario, Sandra G.F. Bukkens and David Pimentel (1997), 'Linking Technology, Natural Resources, and the Socioeconomic Structure of Human Society: Examples and Applications', *Advances in Human Ecology*, **6**, 131–200.

Giampietro, Mario, Giovanni Cerretelli and David Pimentel (1992), 'Energy Analysis of Agricultural Ecosystem Management: Human Return and Sustainability', *Agriculture, Ecosystems & Environment*, **38**, 219–44.

Grübler, Arnulf (1998), *Technology and Global Change*, Cambridge: Cambridge University Press.

Haberl, Helmut (1997), 'Human Appropriation of Net Primary Production as an Environmental Indicator: Implications for Sustainable Development', *Ambio*, **26**(3), 143–6.

Haberl, Helmut (2001), 'The Energetic Metabolism of Societies, Part I: Accounting Concepts', *Journal of Industrial Ecology*, **5**(1), 11–33.

Haberl, Helmut and Fridolin Krausmann (2001), 'Changes in Population, Affluence and Environmental Pressures During Industrialization. The Case of Austria 1830–1995', *Population and Environment*, **23**(1), 49–69.

Hall, Charles A.S., Cutler J. Cleveland and Robert K. Kaufmann (1986), *Energy and Resource Quality, The Ecology of the Economic Process*, New York: Wiley Interscience.

Hall, Charles A.S., Carlos L. Perez and Gregoire Leclerc (2000), *Quantifying Sustainable Development, The Future of Tropical Economies*, San Diego, CA: Academic Press.

Hwaletz, Otto (2001), *Die österreichische Montanindustrie im 19. und 20. Jahrhundert*, Wien: Böhlau.

IEA (2004), 'Energy Statistics of OECD Countries' (CDROM version), www.iea.org.

Krausmann, Fridolin (2001), 'Land Use and Industrial Modernization: An Empirical Analysis of Human Influence on the Functioning of Ecosystems in Austria 1830–1995', *Land Use Policy*, **18**(1), 17–26.

Krausmann, Fridolin (2003), 'The Transformation of European Land Use Systems During Industrial Modernization. A Biophysical Analysis of the Development in Austrian Villages Since 1830', in T.I. Lyubina et al. (eds), *Regional Tendencies of the Interaction of Human Beings and Nature in the Process of the Transition from the Agrarian to Industrial Society*, Tver: Ministry of Education of the Russian Federation, pp. 187–215.

Krausmann, Fridolin (2004), 'Milk, Manure and Muscular Power. Livestock and the Industrialization of Agriculture', *Human Ecology*, **32**(6), 735–73.

Krausmann, Fridolin (2006a), 'Land Use and Socio-economic Metabolism in Pre-industrial Agricultural Systems: Four 19th-century Austrian Villages in Comparison', *Social Ecology Working Paper 72*, Institute of Social Ecology: Vienna.

Krausmann, Fridolin (2006b), 'Landnutzung und Energie in Österreich 1750 bis 2000', in Rolf-Peter Sieferle et al. (eds), *Das Ende der Fläche. Zum Sozialen Metabolismus der Industrialisierung*, Wien: Böhlau.

Krausmann, Fridolin and Helmut Haberl (2002), 'The Process of Industrialization from the Perspective of Energetic Metabolism. Socioeconomic Energy Flows in Austria 1830–1995', *Ecological Economics*, **41**(2), 177–201.

Krausmann, Fridolin, Heinz Schandl and Niels B. Schulz (2003), *Vergleichende Untersuchung zur langfristigen Entwicklung von gesellschaftlichem Stoffwechsel und Landnutzung in Österreich und dem Vereinigten Königreich*, Stuttgart: Breuninger Stiftung.

Krausmann, Fridolin, Helmut Haberl, Karl-Heinz Erb and Mathis Wackernagel (2004), 'Resource Flows and Land Use in Austria 1950–2000: Using the MEFA Framework to Monitor Society–Nature Interaction for Sustainability', *Land Use Policy*, 21(3), 215–30.

Krausmann, Fridolin, Helmut Haberl, Niels B. Schulz, Karl-Heinz Erb, Ekkehard Darge and Veronika Gaube (2003), 'Land-use Change and Socioeconomic Metabolism in Austria. Part I: Driving Forces of Land-use Change: 1950–1995', *Land Use Policy*, 20(1), 1–20.

Leach, Gerald (1976), *Energy and Food Production*, Guildford: IPC Science and Technology Press.

Maddison, Angus (2001), *The World Economy. A Millennial Perspective*, Paris: OECD.

McNeill, John R. (2000), *Something New Under the Sun. An Environmental History of the Twentieth Century*, London: Allen Lane.

ÖSTAT (ed.) (1979), *Geschichte und Ergebnisse der zentralen amtlichen Statistik in Österreich 1829–1979. Festschrift aus Anlass des 150jährigen Bestehens der zentralen amtlichen Statistk in Österreich*, Wien: Kommissionsverlag.

Peterseil, Johannes, Thomas Wrbka, Christoph Plutzar, Ingrid Schmitzberger, Andrea Kiss, Erich Szerencsits, Karl Reiter, Werner Schneider, Franz Suppan and Helmut Beissmann (2004), 'Evaluating the Ecological Sustainability of Austrian Agricultural Landscapes – the SINUS approach', *Land Use Policy*, 21(3), 307–20.

Pimentel, David and Marcia Pimentel (1996), *Food, Energy and Society*, Niwot: University Press of Colorado.

Pimentel, David, Wen Dazhong and Mario Giampietro (1990), 'Technological Changes in Energy Use in U.S. Agricultural Production', in Stephen R. Gliessman (ed.), *Agroecology, Researching the Ecological Basis for Sustainable Agriculture*, New York: Springer, pp. 305–21.

Pimentel, David, L.E. Hurd, A.C. Bellotti, M.J. Forster, I.N. Oka, O.D. Sholes and R.J. Whitman (1973), 'Food Production and the Energy Crisis', *Science*, 182, 443–9.

Polanyi, Karl (1971), *The Great Transformation. Politische und ökonomische Ursprünge von Gesellschaften und Wirtschaftssystemen*, Wien: Europaverlag.

Pomeranz, Kenneth (2000), *The Great Divergence: China, Europe, and the Making of the Modern World Economy*, Princeton, NJ: Princeton University Press.

Schulz, Niels B., Fridolin Krausmann and Helmut Haberl (2001), 'Die Bedeutung der Flächeninanspruchnahme durch Gebäude und Infrastruktur für ökosystemare Prozesse am Beispiel der gesellschaftlichen Aneignung von Nettoprimärproduktion, Österreich 1830–2020', in Umweltbundesamt (ed.), *Versiegelt Österreich? Der Flächenverbrauch und seine Eignung als Indikator für Umweltbeeinträchtigungen*, Wien: Umweltbundesamt, pp. 97–102.

Sieferle, Rolf P. (1997), *Rückblick auf die Natur. Eine Geschichte des Menschen und seiner Umwelt*, München: Luchterhand.

Sieferle, Rolf P. (2001), *The Subterranean Forest. Energy Systems and the Industrial Revolution*, Cambridge: The White Horse Press.

Sieferle, Rolf P., Fridolin Krausmann, Heinz Schandl and Verena Winiwarter (2006), *Das Ende der Fläche: Zum Sozialen Metabolismus der Industrialisierung*, Köln: Böhlau.

Smil, Vaclav (1992), 'Agricultural Energy Costs: National Analysis', in Richard C. Fluck (ed.), *Energy in Farm Production*, Amsterdam, London, New York, Tokyo: Elsevier, pp. 85–100.

Turner, Billie L.I., William C. Clark, Robert W. Kates, John F. Richards, Jessica T. Mathews and William B. Meyer (1990), *The Earth as Transformed by Human Action: Global and Regional Changes in the Biosphere over the Past 300 Years*, Cambridge: Cambridge University Press.

Tüxen, Reinhold (1956), 'Die heutige potentielle natürliche Vegetation als Gegenstand der Vegetationskartierung', *Angewandte Pflanzensoziologie*, **13**, 5–42.

UN (2002), *Energy Statistics Yearbook 1999*, New York: United Nations, Department of Economic and Social Affairs.

Vitousek, Peter M., Paul R. Ehrlich, Anne H. Ehrlich and Pamela A. Matson (1986), 'Human Appropriation of the Products of Photosynthesis', *BioScience*, **36**(6), 363–73.

Wrbka, Thomas, Karl-Heinz Erb, Niels B. Schulz, Johannes Peterseil, C. Hahn and Helmut Haberl (2004), 'Linking Pattern and Processes in Cultural Landscapes. An Empirical Study Based on Spatially Explicit Indicators', *Land Use Policy*, **21**(3), 289–306.

Wright, David H. (1990), 'Human Impacts on the Energy Flow through Natural Ecosystems, and Implications for Species Endangerment', *Ambio*, **19**(4), 189–94.

Wrigley, Edward A. (1988), *Continuity, Chance and Change. The Character of the Industrial Revolution in England*, Cambridge: Cambridge University Press.

3. The fossil-fuel-powered carbon sink: carbon flows and Austria's energetic metabolism in a long-term perspective

Karl-Heinz Erb, Helmut Haberl and Fridolin Krausmann

The accumulation of greenhouse gases, above all, carbon dioxide (CO_2), in the atmosphere is one of the most important driving forces of global environmental change. Humans directly alter the global carbon cycle mainly through two interrelated processes: 1) land use and 2) combustion of carbon-rich materials, above all, fossil fuels. The increased carbon concentration in the atmosphere resulting from these processes, together with rising levels of other greenhouse gases, is commonly thought to induce fundamental changes in the global climate. Climate change not only affects mean temperature and precipitation but is also responsible for increased frequency and severity of extreme events such as droughts, storms or floods.[1] The management of human-induced carbon flows is therefore high on the agenda of global sustainability efforts as exemplified by the negotiation processes aiming at achieving reductions in greenhouse gas emissions that have resulted, among other things, in the Kyoto Protocol. Much research is currently being conducted to quantify these processes on global as well as regional, national or local scales to help deal with this global sustainability problem.

Less attention has, however, been paid so far to the interrelations between land use and fossil-fuel-derived carbon emissions. In this chapter we will argue that these interrelations are indeed important, and that the socioeconomic metabolism approach can be useful in understanding them. Specifically, we will argue that the process of industrialization, powered by fossil fuels and resulting in surging CO_2 emissions from industrial processes, has also brought about land-saving technological innovations in agriculture, thus resulting in increases in forest stocks – that is, in the emergence of a terrestrial carbon sink.

3.1 GLOBAL CARBON FLOWS: SOME BASIC INFORMATION

Terrestrial ecosystems play an important role in the global carbon cycle, as they represent a major natural absorption mechanism for atmospheric carbon. Land use alters the production ecology and hence the carbon absorption of terrestrial ecosystems. It influences not only species composition, dominant growth type, nutrient cycles and many other important aspects of terrestrial ecosystems, but also their net primary production (NPP) and the share of NPP allocated to and stored in various ecosystem components (including roots, stems, foliage, food chains and soil organic carbon). Since carbon makes up about half of the mass of dry matter biomass,[2] all these changes in production ecology are directly relevant for an ecosystem's carbon flows. Moreover, biomass harvest removes carbon from the ecosystem and the further fate of this biomass in the course of socioeconomic metabolism determines whether, in which chemical form (for example, CO_2, CH_4 and others) and how fast this carbon is released into the atmosphere (Gielen 1998).

In many cases, land-use change results in net carbon losses from long-accumulated aboveground or belowground carbon stocks into the atmosphere. Deforestation, for example, greatly reduces the amount of carbon stored in vegetation (in trees in particular) and often also results in large carbon flows from soils to the atmosphere. Changes in land management may, however, also result in a net carbon uptake of ecosystems, for example in the case of reforestation.[3]

The use of carbon-rich fossil fuels such as coal, oil and natural gas results in the emission of carbon in the form of CO_2 and CH_4 through a variety of processes, the most important of which is combustion. Additionally, industrial processes such as cement manufacture contribute some 3–4 per cent to total global industrial CO_2 emissions (Marland, Boden and Andres 2000), however, these will not be dealt with in the remainder of this chapter.

Beyond these forms of direct human impact, climate change and atmospheric carbon concentration itself can alter the carbon balance of ecosystems. For example, an increase in the CO_2 content of the atmosphere can have a fertilization effect, that is, it can induce an increase in biological productivity, which can in turn lead to a build up of carbon stocks. On the other hand, temperature increases may also enhance respiration, which would diminish carbon accumulation or even carbon stocks. Changes in temperature and precipitation might even lead to a loss of tropical forests, which could result in a massive carbon release (Cox et al. 2000). In any case, these forms of indirect human impact on the global carbon balance are at present only partially understood (Melillo et al. 2002) and it is therefore difficult to take them into account when calculating total carbon emissions from and to the atmosphere.

At present, the best available estimates, from the IPCC's Third Assessment Report (Prentice et al. 2003), suggest that in the 1990s, as a consequence of the current elevated CO_2 levels, the oceans absorbed about 1.9 (± 0.6) billion tons of carbon per year (GtC/yr), while terrestrial ecosystems absorbed some 1.4 (± 0.7) GtC/yr. This net carbon absorption of terrestrial ecosystems is the balance of emissions due to land-use change, mainly tropical deforestation, and net carbon absorption in terrestrial ecosystems, mainly in the northern hemisphere, resulting from carbon fertilization.

Figure 3.1 presents a global-level estimate of the magnitude of the two above-mentioned carbon flows directly caused by humans, that is, C emissions caused by land-use change and by fossil fuel combustion. Data were obtained from the Carbon Dioxide Information and Analysis Center (CDIAC). Figure 3.1 shows that carbon emissions related to land use were larger than those caused by fossil fuel use until about 1910. Both flows were of a similar magnitude in the period 1910–50. After World War II, there

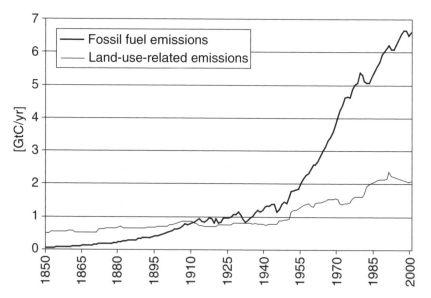

Sources: Fossil-fuel-related emissions: Marland et al. (2000); land-use-related emissions: Houghton (1999), downloads from CDIAC.

Figure 3.1   Global carbon emissions related to land use and fossil fuel use, 1850–2000

was a surge of fossil-fuel-related emissions while land-use-related emissions grew only slowly. Carbon flows related to land-use change rose throughout the whole period but this increase was dwarfed by the surge in fossil-fuel-related carbon emissions after 1950. It is also noteworthy that uncertainties related to fossil fuel emissions are much smaller than those related to the emissions caused by land-use change.

The surge in fossil-fuel-related emissions is a consequence of the global spread of industrialization that was based to a large extent on fossil fuel use, although nuclear power and hydropower were also increasingly exploited during industrialization. Having begun as a historical singularity in England in the 17th and 18th centuries, (largely) fossil-fuel-driven industrialization is currently spreading across the globe (Pomeranz 2000; Sieferle 2001).

The slow growth in land-use-related emissions reflects the process that is still ongoing globally by which types of landcover that store much carbon (that is, above all, forests) are replaced by other kinds of landcover, such as cropland or built-up land, which store little carbon. There are, in any case, significant regional variations to this general trend. Most of the current carbon emissions from land-use change are emerging in the tropics, while the United States and Europe currently have considerable net carbon sinks related to land-use change. For example, the vegetation of the United States absorbed almost 110 megatons of carbon per year (MtC/yr) during the 1990s and Europe's carbon sink was about 18 MtC/yr, while tropical regions emitted almost 2.2 GtC/yr in the same period, mainly due to deforestation (Houghton 1999). It should be noted, however, that these figures are highly contested and uncertain (for example, Caspersen et al. 2000; Houghton, Hackler and Lawrence 1999; Janssens et al. 2003; Pacala et al. 2001).

It is no coincidence that regions that are among those with the highest per-capita emissions of fossil-fuel-related CO_2 have negative or at least small emissions from land-use change, while tropical regions with generally small fossil-fuel-related carbon emissions owing to their small industrial capacities tend to have high land-use-related emissions. This characteristic difference can be explained by the nature of the transition from an agrarian to an industrial energy system. This is demonstrated using 1) general considerations on the role of agriculture and forestry for socioeconomic metabolism, and 2) empirical evidence from Austria, a highly industrialized country with a considerable carbon sink in its vegetation. Although this empirical evidence is less complete than we would wish, we feel that it provides strong support for the thesis we present. Further research would of course be desirable and some ideas on how this could be carried out are also presented here.

3.2   FROM AGRICULTURAL TO INDUSTRIAL METABOLISM: LAND USE CHANGES ITS ROLE

The historian, Rolf Peter Sieferle (2001), has characterized the energy system of agrarian societies as a 'controlled solar energy system'. This notion refers to the fact that biomass – that is, solar energy that has been transformed by green plants (in the process of photosynthesis) into chemically stored energy – is the quantitatively most important source of energy in agrarian societies. 'Controlled' in this context means that agricultural societies mostly use biomass from agro-ecosystems, that is, ecosystems that are to a considerable extent controlled by humans. The energy system of agricultural societies, in other words, is dependent on (and limited by) the net primary production (NPP) of ecosystems accessible and usable for an agrarian community.

Sieferle (2001) contrasts this agrarian energy system with the 'fossil energy system' of industrial economies. This notion does not imply that fossil fuels are used instead of biomass, but rather that area-independent sources of energy become available to an extent sufficient to result in a qualitatively and quantitatively different energy system. This new type of energy system is not constrained by area, because fossil-fuel-dependent or other area-independent prime movers are substituted for human and animal labour (see Chapter 2).

The important aspect here is that during the transition from the agrarian to the industrial regime, the role of land use for socioeconomic metabolism is fundamentally altered: for agrarian societies, land is by far the most important source of energy. This means that agriculture (as a whole, not any individual process) must supply a society with more energy than is invested into it – in other words, agriculture must have a positive 'energy return on investment' or EROI (Hall, Cleveland and Kaufmann 1986; Pimentel et al. 1973). Industrial societies, by contrast, may energetically 'subsidize' agriculture, that is, agriculture may have an EROI of less than 1. Indeed, other criteria, in particular labour efficiency (labour input needed per unit of agricultural produce) and yields per unit area become much more important, often at the expense of energy efficiency (Pimentel, Dazhong and Giampietro 1990; see Chapter 2).

In Austria – as in all the other industrialized countries we have analysed to date – the industrialization of agriculture has resulted in impressive increases in yields, that is, commercial harvest per unit of area cultivated. For example, the commercial yield of many important crops such as wheat, rye, maize or barley has risen by factors of about 5 during the last 50–100 years. This was partly the result of 'improvements' in the utility of crop plants for humans: while in around 1900, grains accounted for only 20–30 per cent of the aboveground biomass of crop plants such as wheat or rye at the time of harvest, this had increased to 50 per cent or more by the 1980s and 1990s. Moreover, the increasing availability of both direct energy (for example, tractors) and indirect energy (for example, fertilizers) for agriculture has greatly increased the NPP of agro-ecosystems. All of this meant that in Austria, for example, about 70 per cent more biomass could be harvested in 1995 than in 1830, using considerably less cropland and less grassland. As a result, despite a sixfold increase in the area covered by settlements, industry and infrastructure, forest area increased quite significantly in the course of industrialization. This was the main reason why the 'human appropriation of aboveground NPP' (aboveground HANPP) fell in Austria between 1830 and 1995 from about 60 per cent to 50 per cent, even though biomass harvest nearly doubled in the same period.

It is important for our argument here to understand that this reduction in HANPP – which, as we shall attempt to demonstrate, has also contributed to the emergence of a considerable carbon sink in Austria – could only be achieved because of the fundamental changes in the role of land use for socioeconomic metabolism outlined above, which were made possible by the use of fossil fuels. In other words, we will demonstrate that Austria's carbon sink is powered by fossil energy and is a 'fossil-fuel-powered carbon sink'.

3.3 CARBON EMISSIONS FROM INDUSTRIAL PROCESSES IN AUSTRIA, 1830–2000

As discussed in detail in Chapter 2, the industrialization of Austria's energy system started in the early 19th century, about 100 years later than in the United Kingdom (Schandl and Schulz 2002, Chapter 4). In 1830, biomass accounted for about 99 per cent of Austria's total energy input, while coal and hydropower each contributed less than 1 per cent. Coal use increased rapidly in the second half of the 19th century, resulting in surging carbon emissions. During World War I, Austria lost its access to rich coal mines in Bohemia, southern Poland and other parts of the former Austro-Hungarian Monarchy, which also contributed to the restructuring of Austria's industry that included a move away from energy-hungry basic industries. The rapid industrial redevelopment of Austria after World War II was fuelled mainly by oil in the first phase – until roughly the mid-1970s – and supplanted by natural gas thereafter. The exploitation of hydropower progressed steadily throughout the 20th century, constituting some 9 per cent of domestic energy consumption in 1995, at which time biomass amounted to about 30 per cent of domestic energy consumption (including nutrition of humans and livestock).[4] In other words, it may be said that Austria completed a process of transition from the agrarian to the industrial mode of subsistence in the period from the early 19th century to the present day.

Figure 3.2 shows the carbon emissions associated with Austria's fossil fuel use during industrial development in the period from 1830 onwards. Note that CO_2 resulting from biomass combustion is not included because biomass is often considered to be 'CO_2 neutral': it is assumed that in burning biomass, the same amount of CO_2 is released into the atmosphere as had been sequestered during plant growth.[5] Moreover, note that we here report only the mass of the carbon contained in flows of carbon-rich materials. In the case of CO_2 emissions – that is, a flow of CO_2 into the atmosphere – the CO_2 emitted weighs about 3.7 times as much as the carbon contained in this flow. In other words, CO_2 emissions of 60 million tons per year are equivalent to carbon emissions of about 16 MtC/yr. The quantity of CO_2 and methane exhaled or otherwise emitted by humans and domesticated animals is excluded from the numbers presented below.

Figure 3.2 shows that CO_2 emissions almost reached half of their current value before World War I, mainly as a result of coal burning. Austria's CO_2 emissions were slightly lower after World War II than they had been before World War I; they quickly doubled to about 16 MtC/yr in the 1970s, and remained roughly constant thereafter. Note that CO_2 emissions per unit of

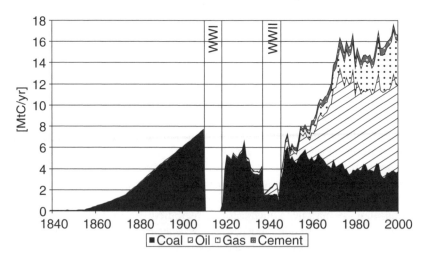

Sources: Coal-related emissions before World War I: Krausmann, Schandl and Schulz (2003). All other data from CDIAC (Marland et al. 2000).

Figure 3.2   The carbon content of Austria's yearly CO_2 emissions from fossil fuel use and cement production 1840–2000

energy are about 30 per cent lower for oil products than for coal. Natural gas combustion releases about 50 per cent less CO_2 than coal per unit of energy. This means that the shift from coal to oil to natural gas also resulted in a significant 'decarbonization' of the energy supply, that is, the increase in CO_2 emissions from fossil fuel combustion was lower than the increase in the amount of energy supplied to Austria's economy in the form of fossil fuel. In addition, hydropower use was expanded (see Chapter 2).

3.4 LAND-USE-RELATED CARBON FLOWS

In this section, we discuss the probable magnitude of the net fluxes of carbon between the atmosphere and terrestrial ecosystems in Austria during the last 150 years. Because large uncertainties exist for many yearly fluxes into and out of terrestrial ecosystems, it is preferable to focus on carbon pools and to estimate average yearly net fluxes by comparing the magnitude of C pools at different points in time. It is important to note that at present, great uncertainty is involved in any such calculation, there-fore the following estimates are no more than a very preliminary assess-ment. Only limited means were available to test the sensitivity of the results for important uncertain input parameters. A comprehensive assessment utilizing all potentially available data sources would require considerably greater resources and would have to use, among others, historical sources such as the Franciscean cadastre as well as many historical statistics that could not be incorporated here. Methods and data used are discussed in Appendix 3.1 at the end of this chapter.

The estimate presented in Figure 3.3 implies that the aboveground com-partment of Austria's terrestrial ecosystems absorbed on average about 0.9 MtC/yr during the period 1880 to 1970. This represents slightly less than half the amount emitted from industrial processes in 1880 and less than 10 per cent of the yearly industrial C emissions in 1970. Carbon uptake, however, was considerably larger in the period from 1965 to 1994: in this period, the average yearly amount of carbon sequestered was about 2.8 MtC/yr or almost one-fifth of the yearly industrial carbon emissions in the period from 1970 to 1995 of about 16 MtC/yr (see above). The increase in aboveground standing crop is the result of two trends: 1) an increase in forest area and 2) an increase in the standing crop of forests per unit area. The increase in forested area con-tributes approximately 20 per cent to the overall increase in carbon stock, with the remainder being due to increases in standing crop per unit area.

With respect to data uncertainty, it can be said that Austria's Federal Environment Agency, which established the time series for Austria's forests from 1960 to 1995, estimates the error of its forest biomass

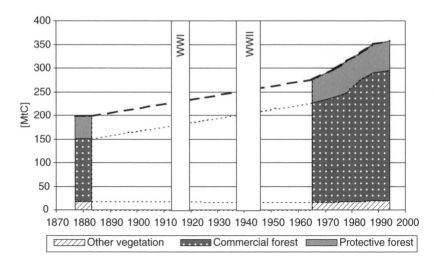

Source: Authors' own calculations. See text for explanation and the Appendix 3.1 for
methods and data sources.

Figure 3.3   *Aboveground standing crop of Austria's terrestrial vegetation,
1880–1994*

inventory to be about 2 per cent. Even if we assume that the value estimated for 1880 has an uncertainty range of 30 per cent – probably an
extreme value – there would still be a marked increase in aboveground
biomass. A conservative estimate with respect to carbon uptake (that is,
the assumption of a high initial value of carbon stocks) would have no
influence on the estimate of yearly carbon uptake in recent decades, but
would result in a considerably lower estimate (0.3 MtC/yr) of carbon
uptake for the period from 1880 to 1970.

It is important to note that the amount of carbon stored in the aboveground compartment of Austria's vegetation in the 1990s is considerably
lower than the amount of carbon Austria's vegetation would store were
land use to be discounted (Erb 2004b). According to Erb's calculation, the
aboveground standing crop of Austria's potential vegetation – that is, the
vegetation that would prevail under current climatic conditions in the
absence of land use – amounts to about 994 MtC or about 2.8 times more
than the current figure of c. 360 MtC. In other words, at current absorption
rates of 2.8 MtC/yr, it would take more than 200 years to reach the hypothetical equilibrium value of potential vegetation. While this is of course
unrealistic because it would imply a complete cessation of land use in
Austria during future centuries, it may still be useful as a hypothetical

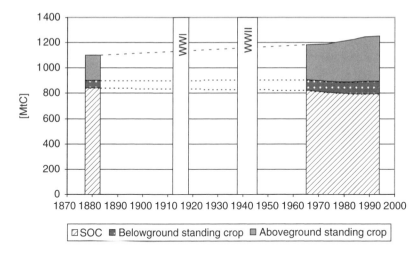

Figure 3.4   *A 'best guess' estimate of the changes in aboveground standing crop (SC_a), belowground standing crop (SC_b) and soil organic carbon (SOC) in Austria, 1880–1995*

assumption to demonstrate the importance of past and present land-use patterns for ecological carbon stocks and flows.

It is very difficult to obtain a more complete picture, however. A rough estimate of the current situation (data refer to 1994) suggests that the amount of carbon contained in the belowground compartment of Austria's terrestrial ecosystems amounts to about 900 MtC (Jonas and Nilsson 2001) or about 2.5 times more than the aboveground carbon stock estimated above. Whether this aboveground trend also holds if changes in the below-ground carbon reservoir are taken into account is an important issue and is discussed in the form of a preliminary assessment in the remainder of this section. See Appendix 3.1 for a discussion of methods and data sources.

Figure 3.4 presents our 'best guess' estimate of the changes in total (above- and belowground) carbon stocks in Austria during the period from 1880 to 1995. According to this estimate, there is an increase in aboveground and belowground standing crop and a small loss in soil organic carbon (SOC). The loss in SOC can be explained by the decrease in extensively managed grasslands (Figure 3.5), which have considerably larger SOC than other agro-ecosystems or even forests. All other landcover categories besides grasslands mainly lose or gain SOC according to changes in their total area. The loss in SOC of extensive grasslands is, according to our 'best guess' calculation, larger than the gain in SOC in forests resulting from the significant increase in forest area during the time period under consideration.

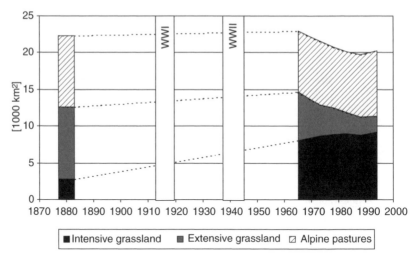

Source: Authors' own calculations, based on Krausmann and Haberl et al. (2003).

*Figure 3.5   Changes in the area of different types of grassland in Austria,
1880–1995*

In order to obtain an initial indication of the uncertainties involved in
our calculations we conducted an analysis of the sensitivity of our results
to changes in the most uncertain of our input parameters, that is, the
assumption on SOC per unit area in the different landcover classes. In this
sensitivity analysis we varied our assumptions on SOC per unit area for all
landcover classes while leaving all other parameters (for example, area of
the landcover classes, aboveground and belowground SC and so on)
unchanged.

To estimate the lower boundary of the carbon sink strength, the lowest
SOC values according to the Austrian Carbon Database (Jonas and
Nilsson 2001) for all landcover classes that increased in area during the
period of analysis and the highest SOC value for all landcover classes that
decreased in area were assumed. In order to obtain an estimate for the
upper boundary of the carbon sink strength, we employed the opposite
assumption. The result of this preliminary sensitivity analysis is displayed
in Figure 3.6.

Figure 3.6 shows that the uncertainties related to assumptions on SOC
per unit area in the landcover classes used in our assessment are not
sufficient to reverse the general trend found in the 'best guess' estimate. This
'best guess' estimate for the overall amount of carbon sequestered each year
remains near 1 MtC/yr for the period from 1880 to 1965 and about

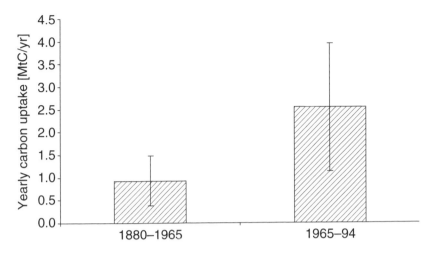

Source: Authors' own calculations.

Figure 3.6   *Estimate of the aggregate (above- and belowground) yearly carbon uptake of Austria's vegetation, 1880–1994*

2.5 MtC/yr for the period from 1965 to 1994 with an uncertainty range of about 0.5 MtC/yr for the earlier period and of almost 1.5 MtC/yr for the later period. This sensitivity analysis does not, however, cover all possible sources of uncertainty in the data, and far more detailed assessments using existing data would be possible if more resources were to be used.[6]

Nevertheless, the result is consistent with the general trends reported above for industrial countries in general and recent assessments for Europe in particular (Janssens et al. 2003). The latter study found that two independent methods (a land-based approach and an inverse modelling method based on measured changes in atmospheric C content) hint at the existence of a considerable carbon sink in Europe's terrestrial biosphere, although both methods did not suffice to rule out completely the possibility that Europe's terrestrial ecosystems could also be a net source of atmospheric carbon.

Our results are in reasonable accordance with the study by Janssens et al. (2003). For example, using a land-based approach such as ours, this study estimates that European forests currently absorb 0.7–1.6 tC per hectare and year, which would translate into a figure of 2.6–6.0 MtC/yr for Austria's forests. Our estimate of the average forest C sequestration from 1965 to 1994 is 2.7 MtC/yr, which is at the lower end of this range. Current studies in general suggest that land-based estimates tend to produce somewhat lower results than inverse modelling methods based on measured changes in atmospheric C content (Janssens et al. 2003; Pacala et al. 2001). It therefore seems more likely than not that our methods underestimate the scale of C sinks on Austria's territory.

To conclude, Austria's terrestrial ecosystems probably constitute a net sink of carbon. This net sink can be explained through two factors. 1) Changes in land use and land cover, above all, an increase in forest area and a reduction in carbon-poor cropland ecosystems. One major land-use-related trend that counters this impact is the conversion of extensively managed grassland to intensively managed grassland, although this process has been by and large completed. 2) Increasing average biomass stocks per unit area in forests, which result from the fact that wood harvest is about 70 per cent lower than wood increment in Austria's forests. This is consistent with studies that found that land-use history and, above all, changes in forest management, account for more than 90 per cent of the C accumulation in forests in the United States (Caspersen et al. 2000).

All these trends are a consequence of the rapid industrialization of Austria's agriculture and of the availability of fossil fuels as a substitute for fuelwood. The industrialization of agriculture proceeded throughout the whole period considered here, but at varying speeds: before the end of World War II, agricultural intensification was slow and fossil-powered technology was not taken up quickly. After World War II, agriculture underwent a process of rapid industrialization, which resulted in overproduction in the 1970s and 1980s, followed by a period during which agricultural policy aimed to reduce overproduction and favoured a diversification strategy instead. Without the introduction of fossil-powered technology on a massive scale, it would have been impossible to achieve the yield gains observed in recent decades that made it possible to reduce farmland area despite the great increase in biomass harvest.

The substitution of fossil fuels for fuelwood started earlier and reached its peak in the early 1970s. Since then, policies have been implemented to foster actively the use of biomass, above all, wood, for space and water heating purposes, both in single household appliances and in small-scale, decentralized district heating systems.

It should be added that the situation outlined here neglects carbon flows associated with Austria's foreign trade. Because Austria is a net importer of fossil fuels there are clearly positive net upstream emissions elsewhere on the globe associated with those fossil fuel imports (for example, CO_2 emissions associated with fuel extraction, fuel processing and fuel transport). The situation regarding biomass is less clear, because the volumes of imported and exported dry matter biomass for Austria are fairly balanced. Calculations of area demand have shown, however, that these biomass trade flows amount to a net import of area (Erb 2004a; Haberl et al. 2003), which hints at the possibility that the carbon emissions 'embedded' in Austria's imports could be larger than those embedded in the exports.

3.5   PUTTING THE PIECES TOGETHER

The results discussed above have demonstrated one obvious and one less obvious point. It is clear that industrialization as experienced to date requires fossil fuels and that, because of this, carbon emissions are related to this process. The increase in these emissions is somewhat mitigated by shifts from carbon-rich to less carbon-intensive fuels, for example, from coal to natural gas, and by the use of more or less carbon-free alternatives, such as hydropower or new renewables (wind power, solar energy and others). The fact that many industrialized countries have terrestrial carbon sinks stemming from increases in forest area and/or biomass stocks in forests has also been noted before (Janssens et al. 2003).

The point that has been less frequently raised, if at all, is that the two processes are systematically interrelated: the availability of fossil fuels as a new, area-independent power source was an indispensable prerequisite for industrialization and it was also a prerequisite for technological changes in agriculture that reduced the amount of area needed per unit of food, feed or other agricultural produce. Together with the possibility of importing considerable amounts of biomass from abroad, these land-saving technological advances are responsible for the fact that forest area is increasing in Austria as well as in many other industrial countries. Moreover, the availability of fossil fuels has reduced the intensity of fuelwood extraction from forests, thus contributing to increases in biomass stocks in forests. Admittedly, we cannot yet unequivocally distinguish between increases in forest biomass stocks caused by changes in forest management and those possibly caused by fertilization effects in terms of increased nitrogen deposition (NO_x or NH_4 emissions), C fertilization or other impacts of climate change, and this gap in understanding should be closed. Present data and knowledge about changes in forest management, however, suggest that at least a large part of the increase in biomass stocks in forests can be accounted for by changes in forest management practices.

A preliminary conclusion from these considerations is that land-use change related to carbon sinks forms part of the 'business as usual' scenario in most industrial countries: even in the absence of any policies directed towards this end, biomass stocks in forests are growing and will probably continue to do so. Nevertheless, industrial countries that have signed the Kyoto Protocol may receive credits for some of these carbon sinks (that is, those associated with so-called ARD – afforestation, reforestation and

Socioecological transitions and global change

deforestation – measures), thus essentially diminishing their targets for cutting fossil-fuel-related emissions. Obviously, these regulations set out in the Kyoto Protocol are of questionable value in terms of mitigating global warming. Concentrating on fossil-fuel-related greenhouse gas (GHG) emissions would have avoided these intricacies and, furthermore, the considerable verification problems associated with the inclusion of land-use-related carbon sources and sinks in international protocols (Jonas and Nilsson 2001; Obersteiner et al. 2000).

A second conclusion is that the reduction of C emissions from fossil fuels is a key element in any solution to the carbon problem, whereas afforestation and other land-use changes in industrialized countries are not. As shown in Figure 3.7, Austria's net carbon emissions are mainly driven by fossil fuel consumption, and the terrestrial carbon sink is a comparatively small side-effect of the transition to fossil fuels and not something that can significantly help to mitigate their adverse environmental impacts.

Third, we may conclude that the common CO_2 neutrality assumption for biomass may be a useful proxy in some cases but is not generally valid. For example, Haberl et al. (2003) have shown that an increase in biomass harvest in Austria that would allow the share of biomass in technical

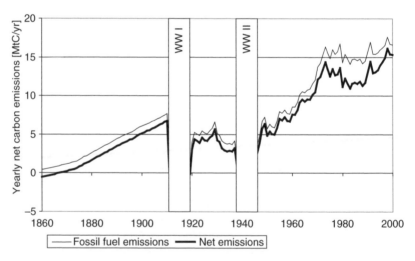

Note: * Constructed by subtracting the authors' 'best guess' estimate for Austria's terrestrial carbon sink from Austria's fossil-fuel-related carbon emissions (Figure 3.2) Sink estimates were derived from mean stock changes as shown in Figure 3.4 (calculated as a 20-year average) and should only be seen as a preliminary approximation.

Figure 3.7   An approximation of Austria's net carbon emissions, 1860–2000*

primary energy consumption to be raised from 12 per cent (trend scenario) to 15 per cent in 2020 would reduce the amount of carbon sequestered in that year in aboveground vegetation by 0.73 MtC. This would be almost identical to the amount of fossil-fuel-related C emissions saved, based on Austria's current fuel mix. This result was mainly derived from the fact that a considerable increase in wood harvest in forests would be required to procure the amount of biomass needed to make this scenario possible.

The overall conclusion is, therefore, that changes in land use and socio-economic energy-use patterns are strongly interrelated, thus supporting the notion that the overall 'energetic metabolism' of societies and not only the use of technical energy captured in conventional energy balances has to be considered in the quest for more sustainable futures. The challenge is either to 'decarbonize' the socioeconomic energy supply or to reduce energy consumption, or both. To analyse which of these options, or which mix of the two, is socially and economically most appropriate, however, is beyond the scope of this chapter and remains a challenge for future research.

APPENDIX 3.1   METHODS AND DATA SOURCES USED TO ESTIMATE LAND-USE-RELATED CARBON STOCKS AND FLOWS

The Aboveground Compartment

It is possible to obtain relatively good estimates with respect to aboveground standing crop, that is, the amount of carbon stored in the biomass present in the aboveground compartment of terrestrial ecosystems. Forests contribute about 95 per cent to the total aboveground standing crop. Therefore, even though the following calculations also include other ecosystems, the results are mainly driven by two parameters: 1) forest area; 2) standing crop per unit area. The latter is strongly dependent on the age structure of forests and also on their species composition.

The assessment of the aboveground carbon stock was based on an accounting approach, using historic landcover data from 1880 to 1995 (Krausmann 2001) and combining them with typical (characteristic) values for carbon storage of each of those landcover types. The area-related values were derived from literature reviews and model assumptions.

Concerning cropland, we assumed standing crop to be equal to aboveground NPP minus herbivory grazing losses, thus calculating peak biomass (that is, biomass stock at the time of harvest). Peak biomass was calculated by multiplying agricultural harvest data taken from statistical sources (for example, ÖSTAT 1992) with appropriate harvest factors taken from the literature (for a detailed description of sources used, see Krausmann 2001). For grasslands (pastures, meadows, alpine pastures and alpine tundra), appropriate factors were derived from literature surveys (for reference, see Erb 2004b). We used the following values: 0.25 kg C/m^2 for pastures and meadows, 0.17 kg C/m^2 for fallow areas and 0.19 kg C/m^2 for alpine pastures at elevations over 1700 m above sea level.

As mentioned above, forests deserve (and have received) most attention. For the period from 1965 to 1994, the assessment was based on forest inventories for the following periods: 1961/70, 1971/75, 1976/80, 1981/85, 1986/90 and 1992/96 (compiled by Weiss et al. 2000). For the year 1880, the assessment of the carbon stocks in Austria's forest was based on a historic forest inventory that contains data for forest stands belonging to national and religious property (Schindler 1885). This inventory contains data on forested area, timber stock, increment and harvest but covers only 10 per cent of the total forest area (between 4 per cent and 53 per cent, depending on the province) of that time. However, qualitative comparisons of national and private forests (Wessely 1853) indicate that the differences with regard to timber stock may be not very large. Hence, and given the lack of alternative datasets, we used these inventory data to assess forest wood stocks. As these forest inventories contain only data on bole wood including bark, we used expansion factors taken from the literature (Burschel, Kürsten and Larson 1993; Cannell 1982; Körner et al. 1993) to obtain estimates of total aboveground standing crop. These expansion factors, reflecting branches and twigs, leaves, fruit, blossoms and understorey, have been developed independently for seven different stand age classes (Mitscherlich 1975; Paulsen 1995). Vegetation gaps, clear-cut areas and bush areas were also considered (Dörflinger et al. 1994; Sattler 1990). Expansion factors were developed for the inventory period 1992/96 and, as a first proxy, have been applied to the other inventories.

The Belowground Compartment

We calculated belowground carbon stocks using a vegetation approach (Paulsen 1995), that is, by assuming a typical carbon content of the soil according to vegetation cover and applying these values to the abovementioned landcover dataset. In order to better reflect carbon content of grasslands – a significant item with regard to the carbon cycle (Ajtay, Ketner and Duvigneaud 1979; Scholes and Hall 1996; Scurlock and Hall 1998) – the landcover dataset was refined by differentiating between extensively and intensively used pastures and meadows. We discerned two compartments for each landcover type: belowground standing crop (for example, roots) and soil organic carbon (SOC), that is, the carbon contained in soils not part of living plants. Belowground standing crop was appraised on the basis of data on the proportion of belowground standing crop in relation to total standing crop using data and assumptions from Körner et al. (1993). For forests, these proportions were developed independently for seven stand age classes and six main forest types (spruce, fir, beech, oak, alluvial forests and open forest). SOC values for the six main forest types were also taken from this source. Values for SOC on non-forested areas were taken from the Austrian Carbon Database (Jonas and Nilsson 2001) for a soil depth of 50 cm. For all SOC values, mean values as well as the uncertainty range were calculated.

ACKNOWLEDGEMENTS

This chapter draws upon research conducted in a large number of empirical studies on the interrelation of land use, land-use change and socioeconomic metabolism. Major parts of the research were conducted as part of the project 'Carbon Household and Socioeconomic Change' (IGBP-26), funded under the Global Change Programme at the Austrian Academy of Science and as part of several projects under the aegis of 'Austrian Landscape Research' (http://www.klf.at/), the research programme of the Austrian Ministry of Education, Science and Culture. Additional funding from the Austrian Science Funds (FWF) projects, 'The Transformation of Society's Natural Relations' (Project No. P16759) and 'Global Human Appropriation of NPP, 1700–2000' (Project No. P16692), is gratefully acknowledged. These projects were part of the overarching 'Land-use Change and Socio-Economic Metabolism' (LUCMETAB) project endorsed by the international LUCC programme.

Empirical findings, methodological aspects and analyses of the data have been published in a number of books, journal articles and reports, in particular, Erb (2004b); Krausmann, Schandl and Schulz (2003); Krausmann and Haberl (2002); Haberl (2001b); Krausmann (2001); Erb et al. (2005) and the forthcoming article Gingrich et al. (2007). Data presented in the text without further reference were first published in one of these publications.

The authors would like to thank Veronika Gaube and Simone Gingrich for providing empirical material and for their valuable cooperation and Marina Fischer-Kowalski for her useful comments at various stages of the project.

78   *Socioecological transitions and global change*

NOTES

1. Houghton (1995); Houghton and Skole (1990); IPCC (2001); Steffen et al. (2002); Watson and Noble (2002).
2. For more details see Körner, Schilcher and Pelaez-Riedl (1993); Lamlom and Savidge (2003); Larcher (1984); Schulze (2000).
3. For further details see Kauppi et al. 1992; Myneni et al. 2001; Schimel et al. 2001; Sedjo 1992.
4. Note that all these figures refer to the 'energetic metabolism' approach (Haberl 2001a, Chapter 2), which differs from conventional energy balances in that it accounts for all socioeconomic energy inputs, not only those used for machinery. Biomass and hydropower each contribute 10–15 per cent to 'technical primary energy use' as accounted for in conventional energy balances.
5. For detailed analyses of such assumptions see Schlamadinger and Marland (1996; 1999).
6. For example, we had no data on organic soil wetlands such as peatlands. These ecosystems can act as considerable carbon sinks and their loss through draining and peat extraction releases considerable amounts of carbon. No data were available on carbon contained in economic imports and exports, C sequestered in enduring wood products, landfills, and so on.

REFERENCES

Ajtay, G.L., P. Ketner and P. Duvigneaud (1979), 'Terrestrial Primary Production and Phytomass', in Bert Bolin, E.T. Degens, S. Kempe and P. Ketner (eds), *The Global Carbon Cycle*, Chichester, New York, Brisbane, Toronto: John Wiley & Sons, pp. 129–82.
Burschel, P., Ernst Kürsten and B.C. Larson (1993), *Die Rolle von Wald- und Forstwirtschaft im Kohlenstoffhaushalt, Eine Betrachtung für die Bundesrepublik Deutschland*, Munich: Forstwirtschaftliche Fakultät der Universität München.
Cannell, Melvin G.R. (1982), *World Forest Biomass and Primary Production Data*, London: Academic Press.
Caspersen, John P., Stephen W. Pacala, Jennifer C. Jenkins, George C. Hurtt, Paul R. Moorcroft and Richard A. Birdsey (2000), 'Contributions of Land-use History to Carbon Accumulation in U.S. Forests', *Science*, **290**, 1148–51.
Cox, Peter M., Richard A. Betts, Chris D. Jones, Steven A. Spall and Ian J. Totterdell (2000), 'Acceleration of Global Warming Due to Carbon-cycle Feedbacks in a Coupled Climate Model', *Nature*, **408**(6809), 184–7.
Dörflinger, Alexander N., Peter Hietz, Rudolf Maier, Wolfgang Punz and Klaus Fussenegger (1994), *Ökosystem Grossstadt Wien, Quantifizierung des Energie-, Kohlenstoff- und Wasserhaushalts unter besonderer Berücksichtigung der Vegetation*, Vienna: Institute of Plant Physiology, University of Vienna.
Erb, Karl-Heinz (2004a), 'Actual Land Demand of Austria 1926–2000: A Variation on Ecological Footprint Assessments', *Land Use Policy*, **21**(3), 247–59.
Erb, Karl-Heinz (2004b), 'Land-use-related Changes in Aboveground Carbon Stocks of Austria's Terrestrial Ecosystems', *Ecosystems*, **7**(5), 563–72.
Erb, Karl-Heinz, Veronika Gaube, Simone Gingrich, Helmut Haberl and Fridolin Krausmann (2005), *Kohlenstoffhaushalt und gesellschaftlicher Wandel, Zusammenhänge von Kohlenstoffflüssen und sozioökonomischer Dynamik in Österreich 1830–2000*, Vienna: Institute of Social Ecology, Klagenfurt University.

Gielen, Dolf (1998), 'Western European Materials as Sources and Sinks of CO_2, A Materials Flow Analysis Perspective', *Journal of Industrial Ecology*, **2**(2), 43–62.

Gingrich, Simone, Karl-Heinz Erb, Fridolin Krausmann, Veronika Gaube and Helmut Haberl (2007), 'Long-term dynamics of terrestrial carbon stocks in Austria. A comprehensive assessment of the time period from 1830 to 2000', *Regional Environmental Change*, **7**(1), 37–47.

Haberl, Helmut (2001a), 'The Energetic Metabolism of Societies, Part I: Accounting Concepts', *Journal of Industrial Ecology*, **5**(1), 11–33.

Haberl, Helmut (2001b), 'The Energetic Metabolism of Societies, Part II: Empirical Examples', *Journal of Industrial Ecology*, **5**(2), 71–88.

Haberl, Helmut (2006), 'The Global Socioeconomic Energetic Metabolism as a Sustainability Problem', *Energy – The International Journal*, **31**(1), 87–99.

Haberl, Helmut, Karl-Heinz Erb, Fridolin Krausmann, Heidi Adensam and Niels B. Schulz (2003), 'Land-use Change and Socioeconomic Metabolism in Austria. Part II: Land-use Scenarios for 2020', *Land Use Policy*, **20**(1), 21–39.

Hall, Charles A.S., Cutler J. Cleveland and Robert K. Kaufmann (1986), *Energy and Resource Quality, The Ecology of the Economic Process*, New York: Wiley Interscience.

Houghton, Richard A. (1995), 'Land-use Change and the Carbon Cycle', *Global Change Biology*, **1**, 275–87.

Houghton, Richard A. (1999), 'The Annual Net Flux of Carbon to the Atmosphere from Changes in Land Use 1850–1990', *Tellus*, **51B**(2), 298–313.

Houghton, Richard A. and David L. Skole (1990), 'Carbon', in Billie L.I. Turner, William C. Clark, Robert W. Kates, John F. Richards, Jessica T. Mathews and William B. Meyer (eds), *The Earth as Transformed by Human Action, Global and Regional Changes in the Biosphere over the Past 300 Years*, Cambridge: Cambridge University Press, pp. 393–408.

Houghton, Richard A., Joseph L. Hackler and K.T. Lawrence (1999), 'The U.S. Carbon Budget: Contributions from Land-use Change', *Science*, **285**, 574–8.

IPCC (2001), *Climate Change 2001: The Scientific Basis*, Cambridge: Cambridge University Press.

Janssens, Ivan A., Annette Freibauer, Philippe Ciais, Pete Smith, Gert-Jan Nabuurs, Gerd Folberth, Bernhard Schlamadinger, Ronald W.A. Hutjes, Reinhart Ceulemans, Ernst D. Schulze, Riccardo Valentini and A.J. Dolman (2003), 'Europe's Terrestrial Biosphere Absorbs 7 to 12% of European Anthropogenic CO_2 Emissions', *Science*, **300**(5625), 1538–42.

Jonas, Matthias and Sten Nilsson (2001), *The Austrian Carbon Database (ACDb). Study – Overview*, Laxenburg: International Institute for Applied Systems Analysis (IIASA), Report i-130.

Kauppi, Pekka E., Kari Mielikäinen and Kullervo Kuusela (1992), 'Biomass and Carbon Budget of European Forests, 1971 to 1990', *Science*, **256**, 70–74.

Körner, Ch., B. Schilcher and S. Pelaez-Riedl (1993), 'Vegetation und Treibhausproblematik: Eine Beurteilung der Situation in Österreich unter besonderer Berücksichtigung der Kohlenstoffbilanz', in Österreichische Akademie der Wissenschaften, Kommission für Reinhaltung der Luft (eds), *Anthropogene Klimaänderung: mögliche Auswirkungen auf Österreich – mögliche Massnahmen in Österreich, Bestandsaufnahme und Dokumentation*, Vienna: Austrian Academy of Sciences, pp. 6.1–6.46.

Krausmann, Fridolin (2001), 'Land Use and Industrial Modernization: An Empirical Analysis of Human Influence on the Functioning of Ecosystems in Austria 1830–1995', *Land Use Policy*, **18**(1), 17–26.

Krausmann, Fridolin and Helmut Haberl (2002), 'The Process of Industrialization from the Perspective of Energetic Metabolism. Socioeconomic Energy Flows in Austria 1830–1995', *Ecological Economics*, **41**(2), 177–201.

Krausmann, Fridolin, Heinz Schandl and Niels B. Schulz (2003), *Vergleichende Untersuchung zur langfristigen Entwicklung von gesellschaftlichem Stoffwechsel und Landnutzung in Österreich und dem Vereinigten Königreich*, Stuttgart: Breuninger Stiftung.

Krausmann, Fridolin, Helmut Haberl, Niels B. Schulz, Karl-Heinz Erb, Ekkehard Darge and Veronika Gaube (2003), 'Land-use Change and Socioeconomic Metabolism in Austria. Part I: Driving Forces of Land-use Change: 1950–1995', *Land Use Policy*, **20**(1), 1–20.

Lamlom, S.H. and R.A. Savidge (2003), 'A Reassessment of Carbon Content in Wood: Variation Within and Between 41 North American Species', *Biomass and Bioenergy*, **25**, 381–8.

Larcher, W. (1984), *Ökologie der Pflanzen*, Stuttgart: Ulmer.

Marland, Gregg, T.A. Boden and R.J. Andres (2000), 'Global, Regional, and National CO_2 Emissions', in CDIAC (eds), *Trends: A Compendium of Data on Global Change*, Oak Ridge, TN, USA: Carbon Dioxide Information Analysis Center (CDIAC), Oak Ridge National Laboratory (ORNL).

Melillo, Jerry M., P.A. Steudler, John D. Aber, K. Newkirk, H. Lux, F.P. Bowles, C. Catricala, A. Magill, T. Ahrens and S. Morrisseau (2002), 'Soil Warming and Carbon-cycle Feedbacks to the Climate System', *Science*, **298**, 2173–6.

Mitscherlich, Gerhard (1975), *Wald, Wachstum und Umwelt*, Frankfurt: Sauerländer.

Myneni, Ranga B., J. Dong, Compton J. Tucker, Robert K. Kaufmann, Pekka E. Kauppi, Jari Liski, L. Zhou, V. Alexandreyev and M.K. Hughes (2001), 'A Large Carbon Sink in the Woody Biomass of Northern Forests', *Proceedings of the National Academy of Sciences*, **98**(26), 14784–9.

Obersteiner, Michael, Yuri Ermoliev, Michael Gluck, Matthias Jonas, Sten Nilsson and Anatoly Z. Shvidenko (2000), *Avoiding a Lemons Market by Including Uncertainty in the Kyoto Protocol: Same Mechanism – Improved Rules*, Laxenburg: International Institute for Applied Systems Analysis (IIASA), Report IR-00-043.

ÖSTAT (1992), *Ergebnisse der landwirtschaftlichen Statistik im Jahre 1991*, Vienna: Österreichisches Statistisches Zentralamt, Beiträge zur landwirtschaftlichen Statistik 1062.

Pacala, Stephen W., George C. Hurtt, D. Baker, P. Peylin, Richard A. Houghton, Richard A. Birdsey, L. Heath, E.T. Sundquist, R.F. Stallard, Philippe Ciais, Paul R. Moorcroft, John P. Caspersen, E. Shevliakova, Berrien Moore III, G. Kohlmaier, Elisabeth A. Holland, M. Gloor, M.E. Harmon, Song-Miao Fan, Jorge L. Sarmiento, Christine L. Goodale, David S. Schimel and Christopher B. Field (2001), 'Consistent Land- and Carbon Atmosphere-based U.S. Carbon Sink Estimates', *Science*, **292**, 2316–20.

Paulsen, J. (1995), *Der biologische Kohlenstoffvorrat der Schweiz*, Zürich: Rüegger.

Pimentel, David, Wen Dazhong and Mario Giampietro (1990), 'Technological Changes in Energy Use in U.S. Agricultural Production', in Stephen R. Gliessman (ed.), *Agroecology, Researching the Ecological Basis for Sustainable Agriculture*, New York: Springer, pp. 305–21.

Pimentel, David, L.E. Hurd, A.C. Bellotti, M.J. Forster, I.N. Oka, O.D. Sholes and R.J. Whitman (1973), 'Food Production and the Energy Crisis', *Science*, **182**, 443–9.

Pomeranz, Kenneth (2000), *The Great Divergence: China, Europe, and the Making of the Modern World Economy*, Princeton, NJ: Princeton University Press.

Prentice, I.C., G.D. Farquhar, M.J.R. Fasham, Michael L. Goulden, Martin Heimann, V.J. Jaramillo, H.S. Kheshgi, C. Le Quéré, Robert J. Scholes and D.W.R. Wallace (2003), 'The Carbon Cycle and Atmospheric Carbon Dioxide', in Richard A. Houghton, Y. Ding, D.J. Griggs, M. Noguer, P.J. van der Linden, X. Dai, K. Maskell and C.A. Johnson (eds), *Climate Change 2001: The Scientific Basis*, Cambridge: Cambridge University Press, pp. 183–237.

Sattler, P. (1990), *Oberirdische Biomasse und Nährelemente von Kahlschlagvegetation in Wieselburg, Pöggstall und Göttweig (NÖ)*, Vienna: Diplomarbeit an der Universität für Bodenkultur.

Schandl, Heinz and Niels B. Schulz (2002), 'Changes in the United Kingdom's Natural Relations in Terms of Society's Metabolism and Land use from 1850 to the Present Day', *Ecological Economics*, **41**(2), 203–21.

Schimel, David S., J.I. House, K.A. Hibbard, Philippe Bousquet, Philippe Ciais, P. Peylin, Bobby H. Braswell, Michael J. Apps, D. Baker, A. Bondeau, Josep Canadell, G. Churkina, Wolfgang Cramer, A.S. Denning, Christopher B. Field, Pierre Friedlingstein, Christine L. Goodale, Martin Heimann, Richard A. Houghton, Jerry M. Melillo, Berrien Moore III, D. Murdiyarso, Ian R. Noble, Stephen W. Pacala, I.C. Prentice, M.R. Raupach, P.J. Payner, Robert J. Scholes, Will Steffen and Christian Wirth (2001), 'Recent Patterns and Mechanisms of Carbon Exchange by Terrestrial Ecosystems', *Nature*, **414**, 169–72.

Schindler, Karl (1885), *Die Forste der in Verwaltung des K.K. Ackerbau-Ministeriums stehenden Staats- und Fondsgüter*, Wien: Druck und Verlag der kaiserlich-königlichen Hof- und Staatsdruckerei.

Schlamadinger, Bernhard and Gregg Marland (1996), 'The Role of Forest and Bioenergy Strategies in the Global Carbon Cycle', *Biomass and Bioenergy*, **10**(5/6), 275–300.

Schlamadinger, Bernhard and Gregg Marland (1999), 'Net Effect of Forest Harvest on CO_2 Emissions to the Atmosphere: A Sensitivity Analysis on the Influence of Time', *Tellus*, **51B**, 314–25.

Scholes, Robert J. and D.O. Hall (1996), 'The Carbon Budget Tropical Savannas, Woodland and Grasslands', in D.O. Hall, A.I. Breymeyer, J.R. Melillo and G.I. Agren (eds), *Global Change: Effects on Coniferous Forests and Grasslands*, Chichester, UK: Wiley, pp. 69–100.

Schulze, Ernst D. (2000), *Carbon and Nitrogen Cycling in European Forest Ecosystems*, Berlin, Heidelberg, New York: Springer.

Scurlock, J.M.O. and D.O. Hall (1998), 'The Global Carbon Sink: A Grassland Perspective', *Global Change Biology*, **4**, 229–33.

Sedjo, R.A. (1992), 'Temperate Forest Ecosystems in the Gobal Carbon Cycle', *Ambio*, **21**(4), 274–7.

Sieferle, Rolf P. (2001), *The Subterranean Forest. Energy Systems and the Industrial Revolution*, Cambridge: The White Horse Press.

Steffen, Will, Jill Jäger, David J. Carson and Clare Bradshaw (2002), *Challenges of a Changing Earth*, Berlin: Springer.

Watson, Robert T. and Ian R. Noble (2002), 'Carbon and the Science–Policy Nexus: The Kyoto Challenge', in Will Steffen, Jill Jäger, David J. Carson and Clare Bradshaw (eds), *Challenges of a Changing Earth*, Berlin: Springer, pp. 57–64.
Weiss, Peter, Karl Schieler, K. Schadauer, Klaus Radunsky and Michael Englisch (2000), *Die Kohlenstoffbilanz des österreichischen Waldes und Betrachtungen zum Kyoto-Protokoll*, Wien: Umweltbundesamt.
Wessely, Joseph (1853), *Die oesterreichischen Alpenlaender und ihre Forste*, Wien: Wilhelm Braumüller.

4. The great transformation: a socio-metabolic reading of the industrialization of the United Kingdom

Heinz Schandl and Fridolin Krausmann

4.1 INTRODUCTION

This chapter deals with the biophysical foundations and consequences of the historical transition from an agrarian to an industrial socioecological regime, which, starting from United Kingdom, the Netherlands and Belgium, encompassed most European countries, the United States and some other countries in the course of the 19th century (Grübler 1994). The perspective on industrialization presented in this chapter focuses on the transformation of the socioeconomic energy system and changes in society–nature interactions and therefore complements and extends classical economic-historical approaches (Fischer-Kowalski and Haberl 1993; Sieferle 2001; Wrigley 1988). From a socioecological perspective, industrialization appears as a process that fundamentally alters the size and structure of socioeconomic metabolism as well as its relation to land use and agriculture.

The chapter explores the specific character of the transition process in the United Kingdom. We will show that industrialization conceived as a socio-metabolic transition is not a continuous and steady process of growth. By using biophysical variables we will distinguish qualitatively different phases of development. The chapter explores the drivers of the metabolic transition and identifies factors that ushered it in during the 17th and 18th centuries.

Our assumptions draw on empirical evidence gathered for the United Kingdom, a forerunner of industrialization. From a historic perspective, the economy in the United Kingdom experienced a transition from a controlled solar energy system to a fully developed industrial nation showing a high level of material and energy use that is characteristic of today's European industrialized countries. In the case of the United Kingdom, this transition took almost four centuries. From a socioecological perspective, it can be characterized by the following key features:

1. A basic requirement of social metabolism is the supply of human nutrition. As a prerequisite to a successful transition, food supply has to accommodate population growth because the basic metabolism of humans has to be secured. The nutritional base is a fundamental element of the socioeconomic energy system and lies at the very core of social metabolism. To allow for a transition from an agrarian mode of production, where the majority of the labour force is engaged in agriculture, to an industrial mode, this basic requirement has to be met. This entails increasing labour productivity in agriculture to free workers from the land and allow for a shift to manufacturing and industry. In the early periods of industrialization this is achieved by an optimization of the traditional agricultural production system within the general conditions of the solar-based energy regime. While the ratio of agrarian to non-agrarian workers before such 'optimization' is at least three to one, afterwards, one agrarian worker feeds five non-agrarian workers.
2. The labour force that leaves agriculture concentrates in urban agglomerations where manufacturers and factories are located. Such concentration of population requires an advanced transport network (such as, for example, the canal system in the United Kingdom) and an energy carrier that is cheap and easy to transport (such as coal). The early use of coal allowed for unprecedented population densities since it overcame the natural limits of energy supply from wood biomass. It allowed for wood to be replaced as the primary source of heating and cooking energy, as well as for process heat in industrial activities. By improving the productivity of brick production, wood could also be replaced as a construction material.
3. Once the problem of conversion of heat into mechanical energy had been solved by the invention of the steam engine, a unique growth mechanism was established. The biophysical basis of the growth mechanism was a technology-resource complex based on coal, iron and the steam engine. Thereby, the limits of the agrarian energy system were fully broken up and physical work was now available to an extent that allowed the United Kingdom to become the most productive and powerful economy on the global scale. The early stages of the coal-based energy regime, though enabling enormous amounts of physical work, did also require a high amount of human labour to fill the gaps within the energy network. This allowed for the employment of rural labour force, which otherwise would have slowed down productivity growth in

the agricultural sector. In this phase of the mature coal-based energy system, the United Kingdom operated at an energetic level far ahead of all other economies and that totally surpassed the limits imposed by the agricultural mode of production, namely, the law of diminishing marginal returns. However, while the industrial system developed quickly, agriculture and also the textile industry remained within the limits of the solar-based energy system, that is, for a long period, two energy regimes operated in parallel.

4. Only by the middle of the 20th century did the fossil-fuel-based energy system take over all social activities including agriculture. This new level of penetration of all social activities by fossil fuel energy was driven by a new technology-resource complex based on oil (later on gas), the internal combustion engine and electricity. This ushered in a new growth phase lasting until the 1980s, where the interlocking of the energy network became ever tighter, replacing most of the solar-based activities in transport and replacing human labour.

4.2 HOW ENGLISH AGRICULTURE BECAME CAPABLE OF PROVIDING THE NUTRITIONAL BASIS FOR POPULATION GROWTH

England had been a backward economy for a long time. In the mid-16th century, England's peripheral location was reflected in low population numbers (and densities) compared with other nations such as France, Germany or Italy. England's population was only a fifth that of France and about a quarter of the German or Italian population. The English economy was weak and advance in agriculture, industry and commerce depended heavily on the import of more sophisticated technologies from the European continent (Wrigley 1988).

Until 1700, the English population had never surpassed five million people. Although estimates for 1300 show a population of already 4–4.5 million people, the plague that arrived in England in 1485 and had several subsequent outbreaks, sharply reduced these numbers to about 2.5 million. In 1600, England had a population of around four million people and the upper estimate for 1300 was only restored in 1615 (Overton and Campbell 1999; Wrigley and Schofield 1981). The increase in population numbers before the 18th century had never been linear; fluctuations were the norm. What is agreed upon today is that after about 1740, the English population entered into an unprecedented growth path. In 1801, the year of the first census, the English population had reached 8.7 million; by 1830, England and Wales had a total of 13.8 million inhabitants (see Table 4.1).

Table 4.1 Population numbers and yearly change in population in England and Wales, Scotland and Ireland

	England & Wales		Scotland		Ireland*	
	(thousands)	(%)	(thousands)	(%)	(thousands)	(%)
1086	2625					
1300	4250	0.2				
1380	2375	−0.7				
1520	2200	−0.1				
1550	2969	1.0				
1600	4110	0.7				
1650	5220	0.5				
1700	5060	−0.1			1905	
1750	5739	0.3	1265		2385	0.5
1801	9061	0.9	1625	0.5	5217	1.6
1831	13994	1.5	2374	1.3	7768	1.3
1851	17983	1.3	2896	1.0	6514	−0.9
1871	22789	1.2	3369	0.8	5398	−0.9
1901	32612	1.2	4479	1.0	4447	−0.6
1931	39988	0.7	4843	0.3	4176	−0.2
1961	46196	0.5	5184	0.2	4246	0.1
1991	50758	0.3	5107	0.0	5138	0.6

Note: * Figures after 1901 refer to an aggregate of the Republic of Ireland and Northern Ireland.

Sources: Mitchell (1994); Wrigley and Schofield (1981).

When the English population started to increase around 1740, this increase was driven by certain exogenous factors reducing child mortality and extending life expectancy (Wrigley and Schofield 1981). In order to meet the requirements of a growing number of people, English agriculture had to adapt to deliver the necessary amounts of food. Applying socioecological thinking, we ask how the agricultural system in England and Wales could be rearranged to support an increasing population in the period from 1740 to 1850, when food trade played an insignificant role and the whole nutritional requirement of the population had to be met by domestic production.

During the 18th century, population grew by around 70 per cent. In the first 30 years of the 19th century, growth increased and amounted to over 50 per cent. Compared with these pronounced growth rates in population, arable land and land for bread grains grew considerably slower (see Table 4.2 below). But because yields for bread grains (wheat and rye) also improved

*Table 4.2   Population and cereals production in England and Wales,
1700 to 1900*

	Population (millions)	Arable Land (km²)	Bread Grains (km²)	Yield (kg/ha)	Production (tons)	Production Per Capita (kg/cap)
1700	5.06	36 000	8 600	1 095	941 700	186
1800	8.66	43 000	11 000	1 470	1 617 000	186
1830	13.28	57 000	14 000	1 868	2 615 200	197
1871	21.50	56 000	13 600	2 000	2 720 000	127
1900	32.25	49 400	24 600	1 883	4 651 010	144

Sources: Collins (2000); Mitchell (1994); Overton and Campbell (1999); *Statistical
Abstract for the United Kingdom* (later *Annual Abstract of Statistics*), 1854ff.

greatly, total production was able to keep pace with population growth. Table
4.2 shows that despite population almost tripling between 1700 and 1830, per
capita supply of cereals remained constant at a level of around 190 kg.[1] From
then on, population growth and domestic agricultural production did not
coincide any more and further growth in population could not be supported
by domestic food production. The nutritional demands of the English population increasingly had to be met by large-scale food imports.

The dramatic growth in food production under the conditions of an agrarian regime is remarkable and is the result of a combination of the expansion
of cropland and intensively used grassland (see Figures 4.1 and 4.2 for
change in land use and arable land between 1800 and 2000) and the optimization of traditional agriculture, based on a complex process of interdependent institutional and technological changes. The English case shows a
very early restructuring of the feudal institutional framework of manorial
lords and peasant communities into a market-oriented system of three agrarian classes, namely, landlords, tenants and agricultural labourers. This new
system of landownership and agricultural wage labour had already taken
shape in the 17th century. As a result, agricultural holdings in England were
larger in size than anywhere else in Europe and the owners of large estates
were more likely to implement technological innovations than were smallholders in other countries. Moreover, capital was available to a degree to
allow for investments to be made in agricultural businesses and also in transport infrastructure and a national market to absorb the agricultural surplus
already formed during the 18th century (Overton 2004).

While the traditional open-field system was based on a defined relationship between husbandry and stock farming, thereby limiting the ability to
specialize, the new enclosed system was market-oriented and rewarded

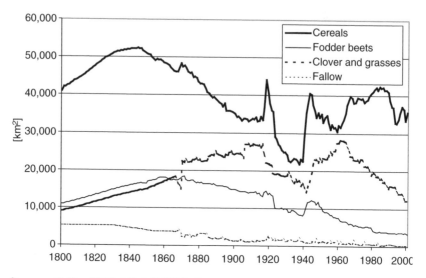

Sources: Collins (2000); Mitchell (1994); Overton and Campbell (1999); *Statistical Abstract for the United Kingdom* (later *Annual Abstract of Statistics*), 1854ff.

Figure 4.1   Land use in the United Kingdom, 1800 to 2000 (km²)

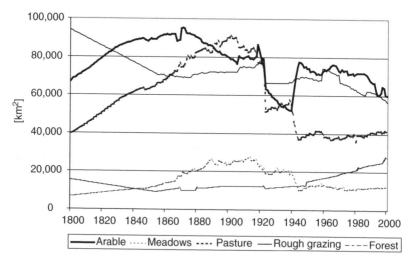

Sources: Collins (2000); Mitchell (1994); Overton and Campbell (1999); *Statistical Abstract for the United Kingdom* (later *Annual Abstract of Statistics*), 1854ff.

Figure 4.2   Arable land in England and Wales, 1800 to 2000 (km²)

specialization and innovation. In the second half of the 17th century, smaller landholders increasingly experienced market pressure. Because their access to markets for loanable funds was restricted, they were less able to innovate and as a consequence, had to sell their property to larger owners (Overton 2004). This resulted in a concentration of landownership mostly driven by the market rather than by oppressive expropriation. Since small businesses were more affected by crop failures, a series of such events further reduced the number of holdings significantly.

Overall, the size of agricultural holdings continuously increased and subsistence peasants as well as small freeholders had to sell their farms and become agricultural labourers. By the end of the 19th century, aggravated competition had reduced the land held by the yeoman freeholders to less than 10 per cent of the total available, while 90 per cent was the property of great landowners (Allen 2000). This process was presumably a precondition for the innovations in agriculture and related productivity gains. This becomes apparent when the situation is compared with continental Europe (France, Poland and Austria), where smallholders and subsistence agriculture dominated land use until the 20th century (Grigg 1980).

The most important technological precondition for improving the productivity of the traditional, solar-based agricultural system was the introduction of new crop rotations replacing the old three-field system. Instead of a fallow year every third year, farmers started to grow fodder crops such as turnips and clover (see Figure 4.1). These new practices of 'Norfolk rotation' increased the availability of livestock fodder enormously and allowed for increases in livestock numbers, which increased both the draught power and the amount of manure for fertilizing the arable land.

Numbers for livestock in agriculture are available from 1812 onwards and show a remarkable increase in the numbers of horses, which were used mainly to provide draught power for ploughing and transport (Figure 4.3). O'Brian and Keyder (1978) have suggested using the ratio of available draught power per unit of arable land as an indicator for agricultural improvements. In 1812, 16.1 gross animal units of draught animals per km^2 of arable land were available in England. In France, according to O'Brian and Keyder, only 8.9 units of horse equivalents were available per km^2, that is, only half of England's capacity. Draught animal availability increased to 17.7 gross animal units per km^2 in 1850 and had a peak of 28.6 in 1905, decreasing thereafter. Horses had an important share in traction in the United Kingdom until the 1940s but then rapidly lost their importance with the mechanization of agriculture. By 1960, animal traction had become negligible and only 1.5 gross animal units of draught animals were available per km^2.

90 *Socioecological transitions and global change*

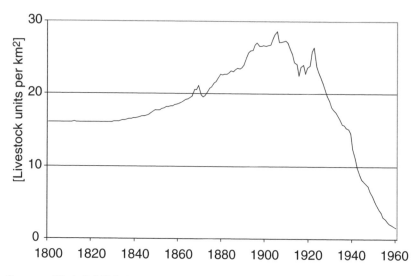

Sources: Mitchell (1994); *Statistical Abstract of the United Kingdom* (later *Annual Abstract of Statistics*), 1854ff.

Figure 4.3 Animal traction (horses and oxen) in the agriculture of the United Kingdom, 1800 to 2000 (LU/km²)

Using gross production of bread grains (Tables 4.2 and 4.3) as a proxy for nutritional supply indicates that the agricultural system in England was able to keep pace with population growth until around 1830. The optimization of agriculture allowed for a doubling of area productivity between 1700 and 1830, but even more significant for the process of industrialization were the achieved increases in agricultural labour productivity: a tripling of food output was accomplished with a rather modest increase in agricultural labour force.[2] In contrast to other European economies, English agriculture very early on achieved a high labour productivity, which was the precondition for supplying a rapidly growing urban-industrial population with sufficient food energy. While in 1600, 70 per cent of the population were still engaged in agricultural activities, in 1871, only 16 per cent of the population could be categorized as agricultural (Overton and Campbell 1999). Employment data available of reasonable quality since the 1831 census allow for estimates of agricultural labourers. Following Taylor (quoted in Collins 1999), in 1840, 1.57 million people were employed in agriculture. However, this number relates only to full-time workers and omits numbers of female workers, farmers' relatives, domestic servants, agriculture-related workers such as gardeners, 'general labourers' and, above all, casual and seasonal workers. Collins (1999) assumes that actual numbers of the agricultural labour force in 1841

Table 4.3   *Population and cereals production in England and Wales, 1700 to 1871 (% change)*

	Population	Arable Land	Bread Grains	Yield	Production	Production Per Capita
1700–1800	71	19	28	34	72	0
1800–30	53	33	27	27	62	6
1830–71	62	−2	−3	7	4	−36
1871–1900	50	−12	81	−6	71	13

Sources: Collins (2000); Mitchell (1994); Overton and Campbell (1999); *Statistical Abstract for the United Kingdom* (later *Annual Abstract of Statistics*), 1854ff.

Table 4.4   *Agricultural productivity in England and Wales*

	Non-agricultural/ Agricultural Population (coefficient)	Arable Land Per Agricultural Worker (ha/worker)	Agricultural Output/ Agricultural Worker (tons/worker)
1700	0.8	3.03	Data unavailable
1801	1.8	2.75	41.6
1831	3.0	3.32	52.1
1871	5.4	3.42	54.8

Sources: Collins (1999); Mitchell (1994); Wrigley and Schofield (1981); *Statistical Abstract of the United Kingdom* (later *Annual Abstract of Statistics*), 1854ff; various census data.

were around 1.79 million, of a total of around seven million economically active people. The number of agricultural workers increased to 1.95 million in 1851 (Collins 1999), but thereafter the numbers declined significantly. In the second half of the 19th century, England began to import food at increasing rates, which had significant impacts on domestic agriculture: from the 1870s onwards, areas of marginal productivity were taken out of production, resulting in a reduction of cropland and agricultural labour force. By 1881, half a million jobs in agriculture had been lost due to further intensification and the closing of agricultural holdings of lower profitability (see Table 4.4 for agricultural productivity in England for 1700–1871).

Notwithstanding the weakness and unreliability of labour data for the agricultural sector, we are able to present several measures of labour productivity that all point to the early advantage of English agriculture compared with continental Europe.

The relative decline in the agricultural labour force was also reflected in early urbanization. By 1800, one-third of the English population was already living in cities with populations larger than 5000 inhabitants. Urban population dramatically increased and reached around 40 per cent by 1830 and 60 per cent by 1871 (Overton and Campbell 1999; Wrigley 1985). Until 1700, urban growth was mostly experienced in London and in important ports, and to a much lesser degree in the historic regional centres. After 1700, urban growth was most pronounced in the new manufacturing towns such as Birmingham, Manchester, Leeds and Sheffield, where population growth by far surpassed that of all other urban agglomerations (Wrigley 1985).

To conclude: what we can see from our analysis is that due to an early rearrangement in landownership structures, availability of capital and a disposition for innovation, the traditional agrarian system in England and Wales did achieve unprecedented improvements in both land and labour productivity. This enabled English agriculture to feed a rapidly growing population in the cities and in the industrial labour force. English agriculture managed stable yields for cereals at 2 tons per hectare, which was far ahead of European standards. Compared with the agricultural system in Central Europe, English yields were twice as high and per capita output was also considerably higher (Turner et al. 2001). The English agricultural regime was not only able to support the population but also yielded monetary surplus to be invested in infrastructure (such as transport) and in other sectors of the economy, predominantly in the textile industry, which had always been closely related to agriculture in terms of raw materials and labour force.

From a social metabolism perspective it is important to emphasize that English agriculture achieved these surprising gains within the limits of the solar-based traditional agrarian system by using all the potential available within the boundaries of this socioecological regime. What also becomes clear is that by 1850, the growth capacity of the solar-based agrarian system was exhausted: as Figure 4.4 shows, no further yield increases could be realized after the 1850s and further growth in population could not be sustained by domestic food production, so that a considerable part of the nutritional requirements, as well as other raw materials, had to be bought on international markets.

In England it was possible to allow the agricultural area (arable and pasture) to increase to an extent where woodlands became nearly eradicated (see Figure 4.2 above), by replacing timber for coal as the dominant source of exosomatic energy. At a certain point in history, around 1870, when the potential of agrarian modernization within the solar-energetic context had been exhausted, the United Kingdom changed its economic strategy by importing increasing quantities of staple food from other world regions (and adjusting for population overspill by migration). This funda-

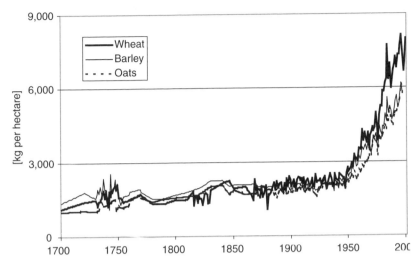

Sources: Mingay (1989); Mitchell (1994); Thirsk (1985); Turner, Beckett and Afton (2001); *Annual Abstract of Statistics*, 1981ff.

Figure 4.4   Crop yields of main cereals in England and Wales, 1700 to 2000 (kg/ha)

mental change in nutritional supply was made possible by the combination of advantages stemming from early industrialization (use of coal and iron) and the superior agricultural system with yields high above the European average both guaranteeing an economic advantage. Based on this economic advantage, the economy of the United Kingdom was able to change its behaviour by making use of its powerful position in the world market.[3]

4.3   INCREASES IN PHYSICAL WORK THROUGH A NEW TECHNOLOGY-RESOURCE COMPLEX: COAL, IRON AND THE STEAM ENGINE

Although coal was known and used in the United Kingdom from ancient times and also during the Middle Ages, its use was restricted to certain places and only small quantities of coal were actually extracted. Around the mid-16th century, yearly extraction was around 0.2 million tons, equivalent to an average of 70 kg per capita. By 1690, annual extraction had already reached 2.9 million tons, equivalent to 350 kg per capita (Nef [1932] 1966). Initially, coal was used for process heat in applications where technical problems were easy to solve, including lime burning, boiling salt and forging metals. For all other manufacturing processes it took some time to develop suitable stoves. By the mid-17th century, technical solutions regarding most of these processes had been developed, except for iron smelting. The iron industry was at that time still using charcoal as the dominant fuel (Hammersley 1973; Riden 1977).

According to Nef ([1932] 1966), in 1700 around 1 million tons of coal were used in industry and manufacturing, while the remaining 1.9 million tons were used in household consumption mainly for cooking and space heating. Based on these assumptions, household consumption would have amounted to an average of around 750 kg per household.[4] Since transport costs affected the possibility of using coal, it was mainly used in areas to which it could be transported on waterways, with only limited use in rural areas unconnected to the river and canal system.

As a first indication of change, coal use for process heat and households already reduced the demand for timber significantly. Timber as a source of primary energy was replaced by coal but, also, bricks replaced construction wood, since they could be produced in larger amounts and at a much lower price when coal was introduced as a fuel in brick kilns.

After the invention of the first steam engine by Newcomen in 1712 and the first successful use of coal in the ironworks of A. Darby in Coalbrookdale (1707–09), coal consumption grew gradually between 1700 and 1770. When the use of coke in ironworks started to spread around 1760 and the more efficient Boulton and Watt steam engine was introduced in 1776, growth in coal consumption became accelerated and was further enforced by the adoption of the puddling-rolling process in 1800 and the Trevithick high pressure steam engine in 1802. By around 1820, despite various technological innovations, per capita coal consumption comprised roughly 1 ton per year.

The real breakthrough was achieved when the steam engine allowed for using coal as a prime mover for the transport of people and goods. The first railway boom from 1835 to 1837 resulted in an average per capita coal consumption of 1.5 tons (Figure 4.5), the second railway boom from 1844 to 1846 contributed to a new level of per capita coal consumption of 2 tons. During the next five decades until 1900, coal consumption reached 4.5 tons per capita and then remained stable at this high level until around 1950, after which coal was gradually replaced by oil and gas.

Figure 4.5 shows that in the early years of fossil fuel energy use, around 1700, two-thirds of coal production were used in household consumption while manufacture and industry had a much smaller share. By 1850, 30 per cent of coal produced was still used domestically but by 1955, the domestic share had been reduced to 15 per cent of total production (Deane and Cole 1967). It should be noted, however, that since coal was increasingly used to produce electricity, part of the electricity current was

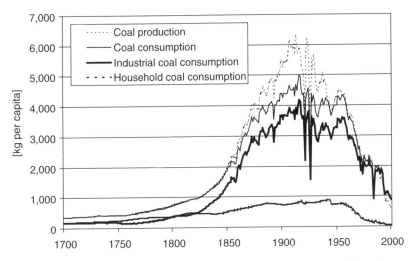

Sources:　Mitchell (1994); Nef [1932](1966); *Statistical Abstract for the United Kingdom* (later *Annual Abstract of Statistics*), 1870ff.

Figure 4.5　*Per capita coal consumption in the United Kingdom's industry and households, 1700 to 2000 (kg/capita)*

used for households. Another fraction of coal was used for coal production itself, enabling water to be pumped out of the coal mining sites. The share of coal use employed by mines themselves remained constant, at between 3 and 6 per cent.

Exports constituted another fraction of coal not available for physical work in the United Kingdom. Foreign trade with coal began around 1850 and became more pronounced around 1890, when 15 per cent of all coal produced was traded internationally. The share of coal destined for export reached around one-third of production in 1910 and remained high until 1940, when exports ceased.

By deducting household coal consumption, consumption in coal mines and exported coal from the production figures, we arrive at the amount of coal productively used in the United Kingdom either for industry, manufacturing or transport (that is, industrial coal consumption as shown in Figure 4.5).

While coal for industrial and manufacturing use remained below the amounts used in households until 1820, industrial coal use experienced sharp growth from this date onwards and reached a peak in the 1920s. Even on a per capita basis, the 1920 level of coal use in industry was eight times that of 100 years earlier, marking a period during which the United Kingdom maintained an advantage over all other economies by operating at a

Socioecological transitions and global change

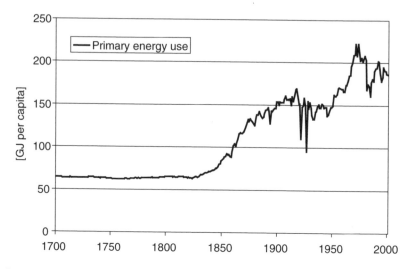

Sources: Mitchell (1994); Nef [1932](1966); *Statistical Abstract for the United Kingdom* (later *Annual Abstract of Statistics*), 1870ff.

Figure 4.6   *Primary energy use in the United Kingdom, 1700 to 2000 (PJ and GJ/capita)*

different, much higher level of energy availability. After World War I, however, the United Kingdom's hegemonic position was eroded and other economic powers, including the United States and Germany, were making grounds, eventually overtaking the United Kingdom in terms of economic leadership (Adams 1982).

Energy use in the United Kingdom across all sources of energy shows a rise in available primary energy (Figure 4.6) and a shift from biotic to fossil fuel energy sources (Figure 4.7). To put it another way, we observe the replacement of the agrarian, solar-based energy system by the fossil-fuel-dominated industrial energy system. Changing from one mode to the other allowed the growth limits of the ancient regime of solar energy use to be overcome and introduced a new growth mechanism, which took different forms during the course of development but was based on the availability of abundant fossil fuel energy resources.

While we can already detect the first indications of a systems change in 1700 when the coal share already constituted 15 per cent of primary energy, this change accelerated around 1800, when timber for energy use was reduced significantly. Coal accounted for 60 per cent of primary energy in 1850 and the energy transition was completed by 1900, when coal accounted for 80 per cent of primary energy use.

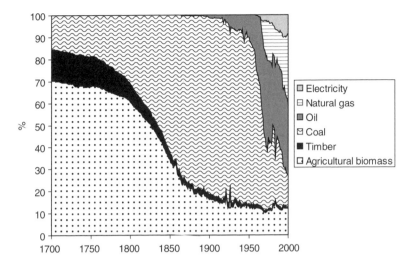

Sources: Mitchell (1994); Nef [1932] (1966); *Statistical Abstract for the United Kingdom* (later *Annual Abstract of Statistics*), 1870ff.

Figure 4.7   *Energy use in the United Kingdom by energy carriers,*
1700 to 2000 (percentage)

Table 4.5   *Primary energy use, final energy and useful energy in the*
United Kingdom, 1700 to 2000 (PJ per year)

	1700 (PJ/yr)	1750 (PJ/yr)	1800 (PJ/yr)	1850 (PJ/yr)	1900 (PJ/yr)	1950 (PJ/yr)	2000 (PJ/yr)
Primary energy use	534	604	1 053	2 264	6 329	7 637	10 950
Final energy	201	240	535	1 608	5 612	6 881	10 001
Useful energy	28	42	111	388	1 527	2 093	4 449

Sources: Mitchell (1994); Nef [1932](1966); *Statistical Abstract for the United Kingdom* (later *Annual Abstract of Statistics*), 1870ff.

Together with the increasing relative importance of coal, the amount of energy use doubled between 1700 and 1800 and again between 1800 and 1850 with a remarkable acceleration in the growth rates (Table 4.5). Beside the increasing ability of the United Kingdom's economy to tap ever higher amounts of fossil fuel energy sources, the efficiency of primary energy used increased significantly, thus boosting the available amount of useful energy still further. Even in phases where primary energy input remained static, the usefulness of energy grew because of technological progress and efficiency gains. This can be assessed by a full energy analysis including primary, final and useful energy.[5]

While a doubling of available primary energy took place in the 18th century (Table 4.5), per capita energy availability increased by only 5 per cent, due to population growth. Nonetheless, the useful energy per capita doubled during this period because of significant efficiency gains in metallurgy, as well as steam engine technology. Possibly, the aggregate rise in the availability of useful energy also implied a slight improvement in the standard of living.

In the 19th century, the availability of primary energy increased in absolute amounts as well as per capita, due to the further implementation of the coal-based fossil fuel energy system. Growth rates, however, were most pronounced in the second half of the 19th century, during which primary energy input grew by a factor of 6. Per capita figures show a doubling of primary energy input, allowing for a new level of primary energy use of 150 GJ per capita per year. This value can be seen as a benchmark for the developed coal-based energy system (see Tables 4.5 and 4.6).

During the same period, useful energy grew by a factor of 14 (factor 5 per capita); efficiency gains were most pronounced in the second half of the 19th century. In the first half of the 20th century, energy availability stagnated and per capita energy consumption decreased. Nevertheless, also in this period the amount of useful energy grew by 40 per cent or 11 per cent per capita. The stagnation was caused by two World Wars and a global economic crisis but it could also be interpreted as a phase of restructuring laying the foundation of the new growth period that started in the 1950s.

When in 1950 the implementation of the oil- and gas-based fossil fuel energy system in combination with more widespread use of electricity took place, pronounced growth rates in absolute and per capita energy use became possible again. By the year 2000, primary energy input was 40 per cent higher

Table 4.6   Primary energy use, final energy and useful energy in the United Kingdom, 1700 to 2000 (GJ per capita per year)

	1700 (GJ/cap/ yr)	1750 (GJ/cap/ yr)	1800 (GJ/cap/ yr)	1850 (GJ/cap/ yr)	1900 (GJ/cap/ yr)	1950 (GJ/cap/ yr)	2000 (GJ/cap/ yr)
Primary energy use	64.6	63.1	68.1	82.2	153.8	151.0	186.7
Final energy	24.3	25.0	34.6	58.4	136.4	136.1	170.5
Useful energy	3.4	4.4	7.2	14.1	37.1	41.4	75.9

Sources: Mitchell (1994); Nef [1932] (1966); *Statistical Abstract for the United Kingdom* (later *Annual Abstract of Statistics*), 1870ff.

than in 1950, resulting in 25 per cent more energy consumption per capita. However, the useful energy tripled (doubled per capita), due to new technologies and new carriers.

Ultimately, the developed fossil fuel energy system differs from the advanced solar-based system by a factor of 3 in terms of available primary energy per capita but by a factor of 20 in terms of available useful energy per capita. Thus, increases in efficiency contributed more to the increase in overall standard of living than did the absolute rise in primary energy consumption.

It is commonly argued that the mechanization of industrial processes increased labour productivity and freed humans from the strains of hard labour. Considering the early phases of fossil-fuel use, this was not typically the case. Nor was this the case for animals. When the railway emerged, many contemporary observers thought that the end of horse-drawn transport had come. It soon became apparent, however, that this was only true for long-distance transport. As the transport volume of goods and also passengers increased rapidly, transportation from and to the railways had to increase accordingly and this transport could only be facilitated by animal traction. Instead of displacing horses, the extension of the railway network led to a drastic increase in draught horses. The penetration of the fossil fuel energy system into the transport sector resulted in an increasing demand for deliveries from human and animal labour, or in another perspective, from the solar-based energy system needed to feed them.

This example can be extended to the industrial labour process. In the early factories, steam power could only contribute to a few, highly concentrated and simple mechanical processes. Steam engines were usually big in order to allow for economies of scale. On a competitive basis, small engines were always out-competed by larger engines. As a result, steam engines were confined to larger factories and could hardly be used in small businesses. In other words, the steam engine area was a driving force for the factory system and the factory system supplied the institutional setting for the new energy carrier-technology complex. Steam engines delivered a highly concentrated mechanical force transmitted to the single working machines via a transmission belt. In this sense, they facilitated the mechanization of all processes requiring such a concentrated force.

The machines had to be operated by workers whose muscular power was needed to fill the gaps between the different mechanical processes and a considerable number of such workers were needed. The early factory system and the steam engine favoured the physically strong industrial worker who had to contribute a multitude of activities, which under existing conditions could not be mechanized. These workers had to perform arduous operations most of the time. Hence, the only real qualification of a worker related to their physical health and strength to be used as a 'human engine'. The share of such workers in the total workforce increased in all industrial countries during the early use of fossil fuels and formed the main type of worker for the entire epoch of the coal-based fossil fuel energy system.

For the seven decades after 1841, coal consumption in industry and the industrial labour force grew in parallel, although the contribution of coal to industrial power was two magnitudes larger than the physical work of humans (Figure 4.8). This indicates that the early coal-based fossil fuel energy system needed ever higher inputs of labour to be able to utilize the power system.

Contemporaries in the 19th and early 20th centuries saw the archetype of the industrial worker as the culminating point of modernity. This was not to last for long: due to technological development and innovation, two processes ensured the gradual disappearance of this type of industrial worker. One process involved the introduction of the electric motor where no scale effects exist. Even a very small electric motor makes sense economically and therefore the cumbersome transmission systems linked to the steam engines could be replaced. Instead of a single and huge steam engine, the factories used numerous much smaller electric engines, which were directly linked to the working machines and obtained electric current

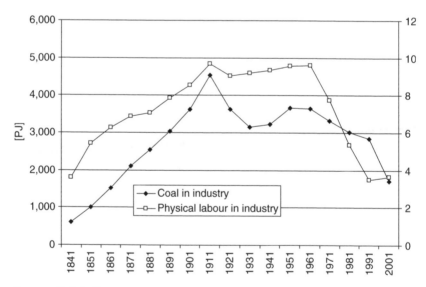

Source: Sieferle, Krausmann et al. (2006).

Figure 4.8   Physical work supplied by labourers and coal in the United Kingdom's industry sector, 1841 to 2001 (PJ)

from a network that was much more closely interconnected than the system of steam power.

With the new driving power of electricity, many steps involved in industrial processing could be mechanized. Human labour was therefore gradually freed from its role in energy conversion and the focus of labour shifted to a new combination of information processing and skilled applications. The requirements concerning the qualification of the workforce changed and the skills of workers became more important. Poorly qualified sectors of the labour force still relying on their mere potential to deliver physical work were increasingly replaced by technical solutions and machinery.

The second process that decoupled production from animate power was the rapid diffusion of the internal combustion engine for transport and draught purposes. The steam-engine-driven railway system could not be extended beyond a certain threshold at which the costs of a further increase would be prohibitive. This problem concerned fixed costs. For passenger transport, the railway system had to be supplemented by horse-drawn carriages or bicycles. The same was true for the transport of raw materials and goods. Also, the number of stations could not be increased significantly, because frequent stops and acceleration would have caused considerable losses in energy. A fundamental breakthrough was achieved when motorized automobiles allowed for an area-wide access for passenger mobility and the transport of goods. Thereafter, all gaps between the railway networks could be filled up, with railways ultimately losing their dominance in passenger and freight transport. The influence of the internal combustion engine and of related technologies had a particularly strong impact in agriculture: within a few decades, the motorized tractor completely replaced draught horses and diminished the need for human labour.

In this process, the network of fossil energy supply was becoming ever narrower, resulting in an almost complete replacement of labourers as human energy converters. The logical end point of such development would be the complete displacement of human labour from the physical production. Although this might sound utopian by today's standards, we should be aware that for 19th-century observers, the perspective of an agrarian labour force falling below 5 per cent of the total workforce appeared just as utopian. Nonetheless, in today's industrial countries around 2.5 per cent of the labour force produces the nutrition for the remaining 97.5 per cent. If cheap energy supply continues, it might be the case that the industrial labourer will deserve the same protection as a relic of an economic past as the peasant farmer enjoys today.

4.4   OVERCOMING THE LIMITS IMPOSED BY THE AVAILABLE LAND AREA

As we have seen, the agrarian solar-based energy regime was characterized by a built-in growth restraint resulting from the limitation of available land. The agrarian socioecological regime is based upon the capacity to tap into flows of solar energy and to modify them to meet its ends. In doing this, there is always room for more effective and efficient use of the available energy resources. Hence, traditional agriculture was able to increase productivity significantly by introducing technical innovations enhancing the yields and allowing for a more efficient use of human labour. Still, agriculture was always confronted with the problem of diminishing marginal returns. In other words, each time yields could be increased significantly the chance for a future increase in yields became less likely, thus moving the system towards saturation. Economists from the 18th and 19th centuries therefore understood the behaviour of the economy as heading towards an unavoidable 'steady state'. Economic growth for long time periods was not conceivable within the limits of an agrarian socio-metabolic regime.

The transition to fossil fuels overcame the limitation imposed by land and the competition of different land-use types (arable, pasture and woodlands) and allowed for unprecedented growth. The 'subterranean forest', which was accessed by the utilization of coal and later oil and gas (Sieferle, Krausmann et al. 2006) made a new and abundant energy source available and disconnected the availability of energy from land use. Contrary to biomass, fossil fuel energy carriers are not area-intensive but constitute stocks that can be mobilized in distinct locations in large quantities.

A simple accounting method allows the relevance of coal to be estimated. We compare the amount of energy provided by coal with a hypothetical forest area that would have been needed to produce the same amount of energy from firewood. Assuming the sustainable yield of forests to be around 5 m^3 of wood per year and hectare (equal to the energy content of 45 GJ), 1 ton of coal (roughly 30 GJ) would correspond to 0.66 hectares of woodland. Respectively, 1 ton of crude oil would correspond to 0.96 hectares and 1 ton of natural gas to 0.78 hectares of woodland (Sieferle, Krausmann et al. 2006).

In order to identify the land area not used for energy production because of the use of coal in Great Britain we have to account for the home consumption, subtract exports from production and add imports. In c. 1700, exports accounted for 2.5 per cent of domestic production of coal and increased steadily to 5 per cent by 1850. Finally, at the beginning of the 20th century, 25 per cent of coal production was exported. This changed abruptly with the outbreak of World War II and coal was never again produced for export. Energy exports from the United Kingdom were then

Table 4.7   Land equivalent of the domestic consumption of fossil energy
             carriers in Great Britain

	Coal (million tons per year)	Crude Oil (million tons per year)	Natural Gas (million tons per year)	Land Equivalents (1000 km²)	Additional Land Area (multiplier of Great Britain's territory)
1700	3			20	0.08
1800	11			76	0.3
1850	53			352	1.5
1875	121			808	3.5
1900	184			1227	5.4
1925	195	7		1371	6.0
1950	206	21		1576	6.9
1975	131	97	58	2259	9.9
2000	54	127	100	2355	10.3

Source: Sieferle, Krausmann et al. 2006.

resumed in 1980, coming from North Sea oil. Exports of fossil fuel energy
carriers, in our analysis, are seen as exports of land equivalents. The
accounting results in the picture presented in Table 4.7.

The woodland area equivalent of fossil fuel energy in the early 18th
century was relatively modest. Three million tons of coal used around 1700
correspond to 20 000 km² of woodland (that is, 8 per cent of the total area
of Great Britain, see Table 4.7). By comparison, 150 years later, fossil fuels
applied as much energy as would the sustainable annual yield of 352 000 km²
of woodland have done. After 1840, the 'subterranean forest' made accessible by the use of coal exceeded the total area of Great Britain. In 1875, coal
provided energy equal to a forest three times the territorial area of Great
Britain. In 1900, a quantity equivalent to woodland covering five times the
area of Great Britain was reached, by 1925, of six times and by 1950, of
nearly seven times that area. While growth rates in coal consumption
decreased during the first half of the 19th century, a new growth impulse was
introduced in the 1950s, when coal was replaced by oil and gas. In energetic
terms, Great Britain today disposes of an area ten times the size of its territory, even if electricity derived from nuclear and water power is disregarded.

These numbers illustrate what a dramatic difference the use of coal makes
in a context in which all competitors still have to rely on their access to land
surface area for energy provision. The historical singularity of the United
Kingdom's development was that a coal economy could be established that

allowed the (already stretched) limitations of the agrarian regime to be surpassed and a path towards long-term rapid economic growth to be opened. The United Kingdom was the first to change from a sporadic use of fossil fuels within the general solar-based energy system to an all-encompassing energy transition in which a new fossil-fuel-based energy regime formed. Such change was made possible by the endowment of the country's territory with rich coal and iron deposits in close proximity to one another, and by a sequence of key technical and industrial innovations.

A central problem of coal mining was the presence of water in the coal pits. Using draught animals to pump water out of the mines was a solution that had its drawbacks: feeding the animals required a substantial land area and agricultural labour force based near to the mines. The coal-fired steam-pumping engine built by Newcomen-Savery (1712), despite its very low efficiency, brought with it a much better solution: it dealt with the water in deeper coal mines and subsequently led to a much higher level of production that could be completely de-linked from land availability. As early as 1733, about 60 Newcomen engines had been produced by the time the master patent expired. The rapidly increasing production of coal powered the industrialization process from this point on.

This transformation had both an energetic and a material dimension. In the agrarian system, the enormous demand for wood in iron and steel production determined a very high price for iron production and thereby a scarcity of supply. Although iron was superior to wood as a construction material and for building tools, it was rarely used. Accordingly, the iron production in Europe in the 17th century was only around 2 kg per capita and year. Iron was used when it could not be replaced by wood, for example, in precision engineering. For other purposes such as ploughs, iron was used for covering the exposed edges of the plough while spades and hoes as well as wheels were usually made from wood. One important user of iron in these early times was the military.

Since the wood demand of ironworks was enormous, they were distributed throughout Europe with important centres of production in Asturia, Styria, north-western France, the Upper Palatinate, the United Kingdom, Sweden and the Urals. The technical upper production limit of an ironworks was 2000 tons per year, determined by the supply of charcoal. Furthermore, ironworks had to be located beside rivers in order to access water power to operate the bellows. This too imposed technical limits on iron production levels.

Changing from charcoal to coke in the production of iron required a series of technical innovations of the metallurgic process, problems that were solved by the development of the puddling-rolling process in the late 18th century. Although the process was not widely adopted until 1800, it was the key breakthrough in iron production and allowed for the large-scale use of wrought iron as a structural material, especially in railway construction during the 1830s and 1840s. Besides using coke, the ironworks used steam-operated bellows to run the blower. The puddling-rolling process was in no sense an isolated invention but was based on a complex innovation process that took place over the course of 100 years and developed solutions for a whole series of related problems. Because of the interconnectedness of these problems, the innovations required to solve them were rather unusual and hence are testament to the particular ingenuity at work in the United Kingdom's development.

The charcoal-based iron production in England in the early 18th century amounted to about 25 000 tons of iron per year. Since the mid-17th century, no substantial increase in production had taken place. It seems that it was economically more profitable to use the area for purposes other than coppicing (to supply wood for the ironworks) and to import iron, in particular from Sweden. Imported iron comprised c. 17 000 tons in 1700 and 40 000 tons in 1750. The per capita supply of iron remained stable at 5 kg/yr.

After 1870, a considerable growth in iron production took place, made possible by the implementation of the new coke-based production process. Pig iron production increased from 40 900 tons in 1770 to 6 million tons a century later, going on to reach 10 million tons before World War I (Figure 4.9). Imports were restricted to precision steel from Sweden, which

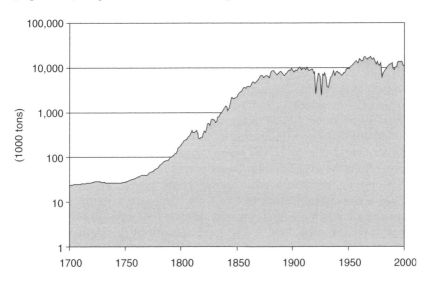

Sources: Hammersley (1973); Mitchell (1994); Riden (1977).

Figure 4.9   Iron production in the United Kingdom, 1700 to 2000 (1000 tons, logarithmic scale)

was produced using charcoal as a fuel. In the early 19th century, exports of iron surpassed imports. Charcoal-based iron production in the mid-18th century required 4000 km² of coppices to produce 40 000 tons of pig iron. Assuming the same technological standard for 1870, iron production would then have needed 600 000 km² of coppices, roughly twice the area of the United Kingdom. Even if we assume efficiency gains in fuel use the amount of land would not have been available and would have introduced a severe bottleneck to production.

4.5   THE SECOND GROWTH PHASE BEGINNING IN THE 1950s

After a period of relative stagnation of energy use at a level of 150 GJ/cap/yr, after World War II, a new growth mechanism was introduced and the per capita energy throughput grew considerably, until the two oil crises of 1973 and 1978 led to a certain stabilization in per capita primary energy use at a level of 200 GJ annually. Most notably, in this period coal was replaced by oil and gas, providing a higher energy density and better transport ability. Coal production in the United Kingdom dropped and coal is today used solely for electricity generation.

The change in the energy regime was far-reaching and established a new technology complex that enforced the use of the fossil fuel energy system. Three key technologies mainly contributed to the ultimate fossil fuel energy transition. Although the basic foundations for their use had been laid in the early 20th century, they could only be fully exploited after World War II in a context of economic restructuring and fast growth. These new technologies include the internal combustion engine, electricity networks and the Haber-Bosch process. Each technology contributed in a very specific way to a self-enforcing process leading to an increase in overall energy use and to a structural change in the energy system, completing the transition process from the previous solar-based energy regime.

The automobile-based road transport of people and commodities created a narrower infrastructure network compared with that of the railways and allowed manufacturing and industry businesses as well as households to be linked up directly. The road network of the United Kingdom has a density of 1200 m/km² compared with c. 80 m/km² in the case of the railway network, that is, the road network density is more than ten times higher than that of the rail network. The increase in road development and car ownership led to the individualization of transport and mobility, resulting in an exponential growth of transport activities (McNeill 2000) and allowing for a new level of spatial differentiation in UK society. Not surprisingly, one-fourth of the energy use in an industrial economy is used in the transport sector and this fraction is still rising.

In this phase of industrial development, for the first time in history, the energy use in private households increased considerably. Besides the automobile, oil- and gas-fired central heating systems, electricity for heating, white appliances (refrigerators, washing machines) and consumer electronics contributed to the increase in household energy consumption. From the 1950s onward, it was most notably the transport and household sector and the rise in the material standard of living that led to the growth in energy consumption in the United Kingdom as well as in all other OECD countries.

The internal combustion engine and the electric motor facilitated an enormous increase in the mechanization of a multitude of labour processes, leading to a substitution of human and animal labour power in the production sector. Working animals disappeared within a short space of time, but human labour in the primary sector and in the industry sectors also decreased markedly. For the first time in history, an increase in industrial production was achieved with a declining number of labour hours. Employment in mining and industry in the United Kingdom reached a peak in 1960, but since then has decreased by 40 per cent (see Figure 4.8). Average working hours of those in employment have been declining since the 1930s (Maddison 2001 and Mitchell 2003).

Finally, the industrialization of agriculture abolished the last growth limit of the old solar-based energy regime. The technological innovations mentioned above coincided with and enabled a tremendous increase in both labour and land productivity in agriculture. Not only did tractor engines replace animal traction and human work, but the labour power per unit of area increased by two orders of magnitude. Around 25 per cent of arable land previously needed for fodder production for traction animals was now made available for food production and woodlands could also grow as pressure on land decreased. Industrial fertilizers solved the problem posed by the limited availability of soil nutrients according to the old agrarian regime. While the fixation of nitrogen by leguminous plants had been a key technology in the agricultural modernization of the 19th century, with application peaking around 1950, this technique was replaced immediately after World War II, leading to a considerable increase in nitrogen availability (that is, amounts increased by a factor of 5).

The introduction of fossil fuel energy in agriculture rendered the necessary combination of different types of land uses and the combination of crop production and livestock-raising obsolete and allowed for a spatial differentiation of agricultural production. At the same time, agriculture changed from being an energy supplier to become a net consumer of energy. The positive return of energy upon invested energy, which was 9:1 in the traditional system, changed to 0.8:1 in today's industrialized agriculture where biomass is produced with a negative energy balance (see Chapter 2 in this volume; Leach 1976; Pimentel and Pimentel 1979).

The new feature of the oil- and gas-based fossil fuel energy regime was the complete enforcement of the new energy regime through all areas of the society, finally changing the character of agriculture and abolishing all limitations that existed under the old socio-metabolic regime. The energy supply was fully decoupled from land and labour. While under the old regime, land management had provided the energetic basis for all other social activities but also implied a growth limit upon these activities, in the new regime this relation has been reversed. Now, the energy sector supplies all areas of society with energy and the economic importance of agriculture has been marginalized.

4.6 CONCLUSIONS: THE THREE PHASES OF THE SOCIOECOLOGICAL TRANSITION

From a socioecological perspective, industrialization appears as a process involving the stepwise decoupling of energy provision from land use and production from animate power. By shifting towards the exploitation of large (fossil fuel) energy stocks rather than tapping renewable but limited solar flows, the transition of the socioeconomic energy system abolishes the constraints on growth imposed by the solar-based agrarian regime. This facilitates unprecedented growth in the size of the physical economy driven by both a surge in population size and in per capita material and energy consumption. Based upon the UK example, we can draw some conclusions on the course of the transition process and the differentiation of periods with specific metabolic characteristics (see also Grübler 1994).

The United Kingdom's socioecological transition from a solar-based agrarian regime to a fossil-fuel-based industrial regime began at some point in the 17th century and gained momentum in the mid-18th century. This early period of the transition process was characterized by the interrelated growth of population and agricultural production: within the space of only 100 years, population density in the United Kingdom increased from a level of 30–35 persons per km^2, typical for many European countries in the 18th century, to a level close to 100 persons per km^2. The growth in population was accompanied by a process involving the optimization of the traditional agricultural production system, commonly referred to as the first agricultural revolution. A combination of institutional and technological changes facilitated unprecedented increases in both area and labour productivity without overstepping the limits of the solar-based energy regime. This enabled the nutritional demands of the growing population to be met but, and this is of crucial significance for the transition process, also the relative increase of the non-agricultural labour force and the development of rapid urbanization: by the beginning of the 19th century, agricultural population accounted for no more than 42 per cent of the United Kingdom's total population. The increases in the energy efficiency of the land-use system were essential for fuelling early industrialization, which was largely based on the use of animate and water power.

But the growth processes during this first period of the energy transition were not only facilitated by a more efficient use of solar flows but were fundamentally related to the increasing exploitation of fossil fuel stocks: only the availability of coal allowed the rapidly growing urban-industrial population to be supplied with energy for space heating and cooking and the rising demand of industry for heat, and later, with the propagation of the steam engine, also for power, to be met. The systematic use of coal represented the first step in de-linking the energy system from land use: coal meant access to an extensive 'subterranean forest' and in turn allowed existing woodland to be abandoned and agricultural area for the provision of food and certain raw materials to be expanded. But most importantly of all, it permitted the mobilization of energy in dimensions unthinkable in terms of solar flows: as early as the 1840s, the area equivalent of the 'subterranean forest' exploited in the United Kingdom exceeded the total land area of the country's territory, indicating an important turning point of the transition process.

A further significant feature of this early period of physical growth is that increases in energy consumption are closely linked to population growth. Per capita consumption remained practically constant until the 1830s at a level of 65–70 GJ/cap/yr (half of which was agricultural biomass).

The early period of industrialization was characterized by the coexistence of the traditional solar-based energy regime with the new fossil-fuel-based industrial regime. In total, by the mid-18th century, the primary energy demand of the United Kingdom had already outgrown the limits of the solar-based agrarian regime. Nevertheless, the provision of food and feed, still the most significant energy sources for work and traction, remained within the old limitations. From the mid-19th century onwards, the tension created by this coexistence rose and agriculture increasingly represented a bottleneck for growth: in the early 19th century the transition of the energy system entered a new phase. In this period, fossil fuels shifted from substituting for firewood to dominating the energy system. With the diffusion of the coal/steam engine/iron/railroad technology complex, the United Kingdom strengthened its position in the world economy and growth of material and energy use accelerated dramatically. From the 1830s on, growth in energy consumption was not merely population-driven, but per capita consumption of energy also began to grow at a fast pace: between 1830 and 1900, energy use doubled from 70 to 150 GJ/cap/yr, a level not reached by any other industrializing economy throughout the 19th century. The high per-capita level, however, was not predominantly due to increases in household energy consumption, but rather to the high level of industrial production, a large fraction of which went into exports. The physical economy exceeded the growth limits imposed by the solar system beyond recall: the 'subterranean forest' made use of in the form of coal expanded in this period to a land-area equivalent of more than five times the size of the United Kingdom.

In contrast to the first phase, coal was now increasingly used to supply power and no longer only heat. However, the advent of the steam engine and the railways did not allow for an absolute de-linking of animate power and production and thus of energy provision and land use: with every newly installed steam engine and growing transport volume by rail, the demand for additional human and animal power also increased and further tightened the link between industrial production and land use. Industrial growth was linked to growing demands for food and feed, yet the optimization of the traditional agricultural production system in the United Kingdom was approaching the limits imposed by the solar-based energy system. From the 1830s onward, cereal yields show hardly any further increase and remain at a level of roughly 2 t/ha/yr, indicating that the potential to further enhance the efficiency of agricultural production was exhausted. In this phase, the fossil-fuel-based energy subsidies for agriculture remained small. By and large, yields and soil fertility had to be maintained through the organization of internal resources and the natural rates of renewal. Despite significant efforts to change this, external inputs into agriculture remained few; mineral fertilizers and steam power were of marginal significance and did not have the potential to improve agricultural production on larger scales. Agriculture remained the bottleneck for growth.

The United Kingdom's 'solution' for this problem was unique in the context of European industrialization: while the industrializers in continental Europe followed suit in improving the efficiency of traditional agriculture throughout the 19th century, the growing nutritional demands of the domestic population of the United Kingdom were met by importing staple foods from newly cultivated lands in the 'New World' and Russia. The import dependency of the United Kingdom with respect to food grew from 15 per cent in the 1850s to more than 60 per cent by around 1900, when more than 25 million UK inhabitants were fed by imports. This process can also be interpreted as one of 'de-linking' energy provision from (domestic) land use: by 1900, the cropland-area equivalent in terms of food imports into the United Kingdom equalled the domestically available cropland area. Clearly, this strategy for overcoming the agrarian bottleneck was not a sustainable and generally applicable blueprint for European industrialization, but the ultimate emancipation of the energy system from land use took another 40 to 50 years to be realized.

After a period of relative stagnation in per capita energy use at a high level of 150 GJ/cap/yr, growth accelerated again after World War II. This new period of growth was closely related to the exploitation of new types of imported fossil fuel energy carriers (oil and gas) and the diffusion of a set of related technologies that opened up new possibilities for the transformation, transportation and use of energy. Electricity and the internal combustion and electric engines, the close-meshed networks of roads and electrification finally increased the general availability of energy and facilitated the area-wide penetration of the fossil-fuel-based energy system into all aspects of life: the dramatic growth in household energy consumption based on new technologies, such as central heating, individual transport and a large number of household appliances, was an historically new phenomenon. The second important cornerstone of the transition process was the detachment of production and transport from animate power: the new fossil-fuel-driven engines were rapidly substituted for animal power in traction and, at an increasing pace, also for human labour in production processes. Finally, a combination of these processes allowed the agricultural bottleneck to be overcome and the energy system to be completely decoupled from land use: mechanization, agrochemicals and transport technologies fundamentally changed the character of the agricultural production system. Direct and indirect external energy inputs were the physical preconditions for further increases in area and labour productivity in agriculture and food output: from the 1940s, the production of edible plant biomass in the United Kingdom grew fourfold in absolute terms and threefold in per capita terms. The price for this surge in output was, however, a decline in energy efficiency: agriculture changed from a source of economically useful energy and an important institution of the energy system into an energy sink. Even though this process was of rather marginal relevance with respect to overall energy consumption, it constituted the final step in the implementation of the fossil-fuel-based energy system and in the de-linking of the energy system and land use.

After World War II, within a short period of only 30 years, the United Kingdom's energy consumption grew by another 60 per cent and per capita energy use reached a level of 200 GJ/cap/yr, which seems typical for most industrialized European countries (Haberl et al. 2006). From our perspective, the post-war period of accelerated European economic and physical growth (the '1950s syndrome', according to Pfister 1996) appears as a distinct but integral step within the long-term process of socioecological transition.

In the 1970s, the so-called oil crises mark a break that might be interpreted as a new turning point in the transition process: the post-war growth period was brought to an abrupt halt, per capita energy use thereafter has oscillated around 200 GJ/cap/yr and overall growth is now driven solely by modest growth rates of population. Triggered by the oil crises, increases in energy prices, the sudden realization of the finite availability of fossil fuels and growing environmental concerns, have driven technological improvements and efficiency gains during the last 30 years. Other factors keeping energy growth in the United Kingdom and Europe at a low pace might be a certain saturation in the diffusion of energy-consuming technologies but also de-industrialization and transfer of energy-intensive and polluting heavy industries to low-income countries outside Europe. It remains unclear whether the current level of energy consumption can be seen as the typical level of industrialized densely populated countries or whether the relative stability of recent decades is merely a temporary interruption of physical growth inherent to industrial societies. It is certain, however, that the current energetic basis of industrial societies is finite and that the fossil-fuel-based industrial society cannot be regarded as a stable socioecological regime.

ACKNOWLEDGEMENTS

Earlier stages of the empirical research on the metabolic transition in the United Kingdom were funded by the German Breuninger Foundation's programme on 'Europe's Special Course into Industrialization' under the leadership of Rolf Peter Sieferle. This final version was completed in the Austrian Science Fund project 'The Historical Transformation of Society's Natural Relations' (Project No. P16759), under the leadership of Marina Fischer-Kowalski.

Parts of the empirical work have been successively presented in a number of publications: Krausmann, Schandl and Schulz (2003); Schandl and Schulz (2000); Schandl and Schulz (2001); Schandl and Schulz (2002); Sieferle, Krausmann et al. (2006).

The authors are grateful to Rolf Peter Sieferle and Verena Winiwarter for providing a stimulating working environment, to Marina Fischer-Kowalski for frequent discussions vital to the progress of our work, as well as to Ted Benton, Peter Dickens, Bob Ayres, Arnulf Grübler, Benjamin Warr, Helmut Haberl, Clemens Grünbühel, Helga Weisz, Karl-Heinz Erb, Reza Nourbakhch-Sabet and Mario Giampietro for their useful comments at various stages of the project.

NOTES

1. How strongly this situation was experienced as a pressure by contemporaries can be judged by the famous pamphlet by Malthus, first published in 1759 (Malthus 1879).
2. Wrigley (1988) identifies the increases in agricultural labour productivity between the 16th and 19th centuries as the single most remarkable feature of the economic history of England.
3. As is well known, the United Kingdom established a number of colonies and dependent regions (in India, Southeast Asia, the United States, Australia and New Zealand) in order to stabilize the privileged access to resources from these countries.
4. Assuming 8.3 million inhabitants in the United Kingdom in 1700 (five million in England and Wales, 2.1 million in Ireland and 1.2 million in Scotland) and an average household size of 4.5 people results in 774 kg of coal per average household and year. With an energy content of 30 megajoules per kilogram (MJ/kg) of hard coal, this amounts to an energy supply for space heating and cooking of urban households of 20–25 gigajoules (GJ/cap/yr) and equals 2–3 m^3 of firewood per capita per year, which is within a plausible range.
5. In order to arrive at the actual usefulness of primary energy within a full energy analysis we apply two steps. First, the primary energy input is allocated to final demand categories of energy users (such as, for example, industry, agriculture, transport) as well as different forms of use (including prime movers, high temperature heat, low and medium temperature heat and lighting). Second, by applying coefficients for the efficiency of final energy use we arrive at useful energy, the actual amount of work delivered (see Sieferle, Krausmann et al. 2006).

REFERENCES

Adams, Richard N. (1982), *Paradoxical Harvest. Energy and Explanation in British History 1870–1914*, Cambridge: Cambridge University Press.

Allen, Robert (2000), 'Agriculture During the Industrial Revolution', in Roderick Floud and Deirdre McCloskey, *The Economic History of Britain since 1700. Volume 1: 1700–1860*, Cambridge: Cambridge University Press.

Collins, Edward J.T. (1999), 'Power Availability and Agricultural Productivity in England and Wales, 1840–1939', in Bas P.J. Van Bavel and Erik Thoen (eds), *Land Productivity and Agro-systems in the North Sea Area (Middle Ages–20th Century). Elements for Comparison*, Turnhout: Brepols Publishers, pp. 209–28.

Collins, Edward J.T. (2000), *The Agrarian History of England and Wales. Volume VII 1850–1914. Part 2*, Cambridge: Cambridge University Press.

Deane, Phyllis and W.A. Cole (1967), *British Economic Growth 1688–1959. Trends and Structure*, Cambridge: Cambridge University Press.

Fischer-Kowalski, Marina and Helmut Haberl (1993), 'Metabolism and Colonization. Modes of Production and the Physical Exchange Between Societies and Nature', *Innovation in Social Science Research*, 6(4), 415–42.

Grigg, David B. (1980), *Population Growth and Agrarian Change. An Historical Perspective*, Cambridge: Cambridge University Press.

Grübler, Arnulf (1994), 'Industrialization as a Historical Phenomenon', in R.H. Socolow et al. (eds), *Industrial Ecology and Global Change*, Cambridge, MA: Cambridge University Press, pp. 43–67.

Haberl, Helmut, Helga Weisz, Christof Amann, A. Bondeau, Nina Eisenmenger, Karl-Heinz Erb and Marina Fischer-Kowalski (2006), 'The Energetic

Metabolism of the EU-15 and the USA. Decadal Energy Input Time-series with Emphasis on Biomass', *Journal of Industrial Ecology*, **10**(4), 151–71.

Hammersley, George (1973), 'The Charcoal Iron Industry and its Fuel 1540–1750', *The Economic History Review. New Series*, **26**(4), 593–613.

Krausmann, Fridolin, Heinz Schandl and Niels Schulz (2003), 'Vergleichende Untersuchung zur langfristigen Entwicklung von gesellschaftlichem Stoffwechsel und Landnutzung in Österreich und dem Vereinigten Königreich', in Rolf Peter Sieferle and Helga Breuninger (eds), *Working Papers of the Breuninger Foundation: The European Special Course. Vol. 11*, Stuttgart.

Leach, G. (1976), *Energy and Food Production*, Guilford: IPC Science and Technology Press.

Maddison, Angus (2001), *The World Economy. A Millennial Perspective*, Paris: OECD.

Malthus, Thomas R. [1759](1879), *Versuch über das Bevölkerungsgesetz*, Berlin: Expedition des Merkur.

McNeill, John R. (2000), *Something New Under the Sun. An Environmental History of the Twentieth-century World*, London: Allen Lane.

Mingay, G.E. (1989), *The Agrarian History of England and Wales. Volume VI. 1750–1850*, Cambridge: Cambridge University Press.

Mitchell, Brian R. (1994), *British Historical Statistics*, Cambridge: Cambridge University Press.

Mitchell, Brian R. (2003), *International Historical Statistics. Europe 1750–2000*, New York: Palgrave Macmillan.

Nef, John U. [1932](1966), *The Rise of the British Coal Industry*, 2 vols., London: Cass.

O'Brian, Patrick and Caglar Keyder (1978), *Economic Growth in Britain and France. Two Paths to the Twentieth Century*, London: George Allen and Unwin.

Overton, Mark (2004), *Agricultural Revolution in England. The Transformation of the Agrarian Economy 1500–1850*, Cambridge: Cambridge University Press.

Overton, Marc and Bruce M.S. Campbell (1999), 'Statistics of Production and Productivity in English Agriculture 1086–1871', in Bas P.J. Van Bavel and Erik Thoen (eds), *Land Productivity and Agro-systems in the North Sea Area (Middle Ages–20th Century). Elements for Comparison*, Turnhout: Brepols Publishers, pp. 189–208.

Pfister, Christian (1996), *Das 1950er Syndrom: Der Weg in die Konsumgesellschaft*, Bern and Vienna: Haupt.

Riden, Philip (1977), 'The Output of the British Iron Industry before 1870', *The Economic History Review. New Series*, **30**(3), 442–59.

Pimentel, David and Marcia Pimentel (1979), *Food, Energy and Society*, London: Edward Arnold.

Schandl, Heinz and Niels Schulz (2000), 'Using Material Flow Accounting to Operationalize the Concept of Society's Metabolism. A Preliminary MFA for the United Kingdom for the Period of 1937–1997', *ISER Working Papers* **2000**(3), Colchester: University of Essex.

Schandl, Heinz and Niels Schulz (2001), 'Eine historische Analyse des materiellen und energetischen Hintergrundes der britischen Ökonomie seit dem frühen 19. Jahrhundert', in Rolf Peter Sieferle and Helga Breuninger (eds), *Working Papers of the Breuninger Foundation: The European Special Course. Vol. 4*, Stuttgart.

Schandl, Heinz and Niels Schulz (2002), 'Changes in the United Kingdom's Natural Relations in Terms of Society's Metabolism and Land Use From 1850 to the Present Day', *Ecological Economics*, **41**(2), 203–21.

Sieferle, R.P. (2001), *The Subterranean Forest. Energy Systems and the Industrial Revolution*, Cambridge: White Horse Press.
Sieferle, R.P., F. Krausmann et al. (2006), *Das Ende der Fläche. Zum gesellschaftlichen Stoffwechsel der Industrialisierung*, Cologne and Weimar: Böhlau Verlag.
Statistical Abstract for the United Kingdom (*Annual Abstract of Statistics*), London: Her Majesty's Stationery Office.
Thirsk, Joan (1985), *The Agrarian History of England and Wales. Volume V. 1640–1750, Part 2*, Cambridge: Cambridge University Press.
Turner, M.E., J.V. Beckett and B. Afton (2001), *Farm Production in England 1700–1914*, Oxford: Oxford University Press.
Wrigley, Edward A. (1985), 'Urban Growth and Agricultural Change: England and the Continent in the Early Modern Period', *Journal of Interdisciplinary History*, **XV**(4), 683–728.
Wrigley, Edward A. (1988), *Continuity, Chance and Change. The Character of the Industrial Revolution in England*, Cambridge: Cambridge University Press.
Wrigley, Edward A. and R.S. Schofield (1981), *The Population History of England 1541–1871. A Reconstruction*, Cambridge: Cambridge University Press.

5. The local base of the historical agrarian – industrial transition and the interaction between scales

Helmut Haberl and Fridolin Krausmann

5.1 INTRODUCTION

Profound changes in Austria's socioecological systems occurred during its transition from an agrarian to the present industrial regime in the years from 1830 to present, as already shown in Chapter 2 of this volume. In this chapter we proceed with an analysis of this transition process in three different local situations. We discuss one urban and two rural case studies and then try to demonstrate a fundamental change in the relations between local and national scales during transitions from an agrarian to an industrial regime.

In the course of the 19th and 20th centuries, Austria was transformed from an advanced agrarian society, in which farmers and their families made up three-quarters of the population, into a classical industrial society, in which manufacturing was dominant, and finally into a modern 'service economy', in which the lion's share of the GDP is produced in service sectors. Overall, total GDP rose almost 30-fold and per-capita GDP 12-fold throughout the period 1830–2000.

This transition was only possible because Austria switched from an area-dependent, biomass-based energy system to an energy system heavily reliant on (mostly imported) fossil fuels. While coal was dominant throughout the 19th century and well into the first half of the 20th century, oil, and later natural gas, gained importance in the second half of the 20th century. These changes in the energy system allowed an almost threefold increase in the amount of primary energy available per capita, and an approximately sevenfold increase in the per-capita availability of useful energy.[1] Because population more than doubled in this period, these increases in per-capita energy turnover implied an approximately sixfold increase in Austria's total primary energy throughput.

These changing patterns in the economy at large, and in the 'energetic metabolism' of society in particular, resulted in massive changes in cultural landscapes and ecosystems, and sustainability problems in general. In 1830, almost all of the energy available to humans in Austria was gained through harnessing the productivity of green plants through agriculture and forestry. This meant that almost all productive area was used either as cropland or as grassland to feed humans and livestock, or as forest to provide timber and woodfuels, and to support livestock through forest grazing and litter extraction. Ecosystems were thus used intensively, probably close to the maximum extent possible with the technology available at that time. Cultural landscapes, however, were highly diverse, and small-scale patterns of settlement, cropland, grassland and forest patches prevailed throughout the country, from the lowlands well into the mountainous regions. Sustainability problems in 1830 were those of an agrarian regime (Haberl et al. 2004a): the challenge was to maintain the often fragile balance between population growth and agricultural production, which might be more accurately described as a balance between the productive capacity of agro-ecosystems that severely constrained energy availability (and thus human and animal populations) on the one hand and the demand for human and animal labour (and therefore sufficiently large population numbers) required for agricultural production on the other (Boserup 1965; Netting 1993). Of course, technology and the organization of production played important roles in this transition process (Boserup 1981; Grübler 1998).

The introduction of fossil fuels alleviated the energetic restrictions experienced in Austria, but as this pattern continues to spread around the globe, new sustainability problems are arising: the dependency on non-renewable resources, including fossil fuels, and the emergence of unsustainable changes in atmospheric composition, in particular, the rising concentration of CO_2 in the atmosphere, which is highly likely to alter the Earth's climate fundamentally (IPCC 2001). While these implications have to be dealt with primarily on the global scale (even though the changes will be felt everywhere on Earth), there have also been important changes in Austria's ecosystems and landscapes. Some of them have been reviewed in Chapter 2, for example, the rising yields on croplands and grasslands, the surging use of mineral fertilizers, the deterioration of agricultural energy efficiency, the shrinking of agricultural areas, the expansion of forests and urban areas, the upward trend of net primary production (NPP) and the reduction in human appropriation of NPP (HANPP) this entailed.

This chapter is largely concerned with changes at local scales in both rural and urban settings, with changes in the relations between urban and rural areas, and with local–national linkages. More generally speaking, these are all issues of spatial organization of both society and ecosystems that are of high relevance for the evolution of cultural landscapes, including their spatial patterns (Peterseil et al. 2004; Wrbka et al. 2004).

The development of transport systems is of course of vital importance for understanding these issues, therefore this chapter places much emphasis on this aspect of technological change.

5.2   TRANSITIONS IN URBAN–RURAL RELATIONS IN TERMS OF ENERGY FLOWS

Chapter 2 has revealed the high significance of changes in society's energetic metabolism for understanding socioecological change during transitions from the agrarian to the industrial regime. Thus we compare here per-capita energy flows in our three case studies, which refer to three very different municipalities, to the national average discussed in Chapter 2. Two local studies, Theyern and Voitsau, concern rural villages in very different ecological as well as socioeconomic settings. Vienna, Austria's capital and largest city by far is the third case study.

Characteristics of the Three Local Cases Studied

Theyern (elevation 250 m a.s.l.) is a mainly cropland farming-dominated lowland community in the Traisen valley near the provincial capital of Lower Austria, St Pölten. Theyern is located in a hilly area less suitable for agriculture than the surrounding lower regions. While it was intensively cropped in the 19th century, agriculture is currently receding. Cropland is being progressively abandoned, while part-time farming and specialized cultures such as orchards and non-agricultural uses are becoming more important. Voitsau (elevation 600–800 m a.s.l.) is located in the granite stock of the 'Waldviertel' area and is characterized by rougher climatic conditions than Theyern. In the early 19th century, grain production in combination with animal husbandry was the dominant type of land use in Voitsau. Current agriculture is dominated by cattle farming (both milk production and fattening) and a mix of cropland and meadows.

These two rural communities are compared with Austria's capital, Vienna. Data for Austria, Theyern and Voitsau refer to approximately 1830 and data for Vienna refer to 1800. No contemporary data are available for the small villages analysed in the historical study. For comparison, we use contemporary data for the municipalities in which the respective village is located, that is, Nussdorf (which includes Theyern) and Kottes (which includes Voitsau). Both are a little more than ten times larger than the villages analysed in the historical study. In the following we refer to Nussdorf/Theyern as a 'lowland system' and to Kottes/Voitsau as an 'upland system'.

Some basic indicators to describe these four systems are displayed in Table 5.1. These figures demonstrate both the huge differences between the systems and the changes between the pre-industrial period studied and the current point in time. It may be interesting to note that, while Austria's population increased by a factor of 2.25 within the period covered, population density remained about constant in the upland system, while its growth was much smaller than the Austrian average in the lowland system (a factor of 1.6). In Vienna, on the other hand, population grew by a factor of 7 – that is, much more rapidly than the Austrian average, indicating the significance of urbanization processes during the transition from the agrarian to the industrial regime. Population density in Vienna was two orders of magnitude larger than in the rural communities (see also Figure 2.1 in Chapter 2).

Table 5.1 shows that agricultural area covered as much as 91 per cent of the area of the upland system in 1830, while the corresponding figure for Austria as a whole was much lower (54 per cent). This can be explained by the fact that a large part of Austria's area is mountainous and hardly suitable for cultivation. The share of agricultural area declined in all three systems. Gross grain yields increased by factors of about 5.5–6.5 in all systems. The aboveground net primary production of agricultural land also increased considerably in all systems. Livestock densities were rather similar in Austria, the lowland and the upland system in 1800/1850, whereas they vary considerably today, with the upland system rearing three times as many animals per hectare of agricultural area than the Austrian average, and the lowland system keeping only one-third of the Austrian average. While ruminants accounted for most of the agricultural animal biomass in 1800/1830, their share had dropped considerably in the Austrian average in 1990/2000, but not in the upland and lowland systems studied.

Energy Flows Around 1830 and 2000: Comparing Local and National Scales

In Table 5.2 we present a comparison of energy flows in these four systems in the early 19th century and the last decade of the 20th century. Data on biomass flows in the upland and the lowland systems and in Austria as a whole are taken from previous work by the authors (Krausmann 2004, 2006a; Krausmann and Haberl 2002). Data on energy flows in Vienna were largely taken from official statistics (see Krausmann 2005).[2]

The results for the lowland and upland systems are compared with the Austrian average obtained from the official Austrian energy balance for 1991 (Alder and Kvapil 1994) in Table 5.2. It shows that total household use of technical final energy per capita calculated for the lowland and

Table 5.1 Basic socioeconomic and agro-ecological indicators of the lowland and upland systems, Vienna and Austria in the 19th and 20th centuries

	1800–30				1990–2000			
	Austria	Lowland	Vienna	Upland	Austria	Lowland	Vienna	Upland
Population [1000 inh.]	3 592	0.10	230	0.13	8 092	1.45	1 618	1.67
Population density [inh./km²]	42	45	4 200	40	96	75	3 899	33
Agricult. population [1000]	2 694	0.10	n.d.	0.13	400	0.54	n.d.	1.10
Agricult. labour force [1000]	1 567	0.07	n.d.	0.09	223	0.23	n.d.	0.46
Agricult. population [%]	75	100	n.d.	100	5	37	n.d.	65
Area [km²]	85 906	2.25	55	3.25	83 400	19	415	51
Agricult. area [km²]	46 627	1.42	n.d.	2.96	33 360	8	79	31
Agricult. area [% of total]	54	63	n.d.	91	40	41	19	61
Farm size [ha agr./farm]	n.d.	8.3	n.d.	11.0	7.2	13.2	n.d.	15.6
Cropland/grassland	0.71	8	n.d.	1.9	9.4	1.9	n.d.	0.8
Gross grain yield [kg/ha]	890	819	n.d.	732	5 708	5 007	n.d.	4 004
Product. agric. land [GJ/ha]	33	38	n.d.	26	84	76	n.d.	114
Livestock density [LU/km²]	17	24	n.d.	24	26	8	n.d.	61
Of which ruminants [%]	95	90	n.d.	96	50	91	n.d.	82

Sources: Krausmann (2004, 2005, 2006a).

Table 5.2 *Per-capita flows of energy [GJ/cap/yr] in the lowland system, the upland system, Vienna and Austria as a whole in the 19th and 20th centuries*

	1800/50				1990/2000			
	Austria	Lowland	Vienna	Upland	Austria	Lowland	Vienna	Upland
DE	*73*	*78*	*0*	*91*	*93*	*76*	*1*	*278*
Biomass	73	78	0	91	61	76	1	278
Fossil fuels	0	0	0	0	15	0	0	0
Hydropower	0	0	0	0	17	0	0	0
Import	*0*	*0*	*35*	*0*	*152*	*77*	*107*	*59*
Biomass	0	0	34	0	23	5	7	8
Fossil fuels	0	0	1	0	124	58	94	41
Electricity	0	0	0	0	5	14	6	10
DEI	*73*	*78*	*35*	*91*	*245*	*153*	*108*	*337*
Biomass	73	78	34	0	84	82	8	286
Fossil fuels	0	0	1	0	139	0	94	0
Hydro/electr.	0	0	0	0	22	0	6	0
Export	*0*	*2*	*0*	*2*	*48*	*36*	*1*	*71*
Biomass	0	2	0	2	24	36	1	71
Fossil fuels	0	0	0	0	19	0	0	0
Electricity	0	0	0	0	5	0	0	0
DEC	*73*	*76*	*35*	*89*	*197*	*117*	*107*	*265*
Biomass	0	76	34	89	60	46	7	215
Fossil fuels	0	0	1	0	120	58	94	41
Hydro/electr.	0	0	0	0	17	14	6	10

Note: DE denotes 'domestic extraction', DEI 'direct energy input' and DEC 'domestic energy consumption'.

Source: See text and Appendix 5.1.

upland systems is similar to total household use of technical final energy according to Austria's official energy balance. Energy use in the lowland system is highest at 51.4 gigajoules per capita per year (GJ/cap/yr) and the upland system uses 50.6 GJ/cap/yr, whereas the Austrian average is 48.5 GJ/cap/yr. Thus, given the uncertainty of such an assessment, we can conclude that household final energy use in the lowland and the upland systems is essentially equal to the Austrian average.

There are, however, meaningful deviations with respect to fuels used. Both rural municipalities, the upland and the lowland systems, use much more biomass (mostly wood) for space and water heating than the Austrian average: in these municipalities, biomass accounts for almost one-third of technical final energy use in the lowland system and about half of technical final energy use in the upland system, and both values are far above the Austrian average of 14 per cent. District heating is not available in both municipalities.[3] Moreover, the upland system was not connected to a natural gas grid in 1991; gas heating systems in the upland system used liquid gas. In the upland system, biomass accounts for almost all (87 per cent) of energy used for space heating.

These figures show that locally available biomass can still play a significant role in technical energy supply in rural areas, while its importance is much lower in urban settings and thus also in the national average, since a large proportion of Austria's population today lives in urban areas. The use of oil products was higher in the Austrian average, due to the higher share of heating oil in space heating compared with the two rural municipalities (upland and lowland), where a high percentage of dwellings used wood as a heating fuel.

Rural–Urban Interrelations: Growing Connectedness and Interdependency

The data contained in Table 5.2 are presented in comprehensive form in Figure 5.1. Quite clearly, in 1800/1830, Austria as a whole resembled the two rural cases while Vienna, as a city, differed completely. Just like Austria as a whole, rural villages were almost self-contained systems, which received little quantifiable inputs from outside (there were a few, but our assessments suggest that these were quantitatively negligible). Unlike Austria, however, both rural systems exported biomass. Even though this export was small in biophysical terms – around 2 GJ/cap/yr for both villages – it was highly valuable and decisive for the functioning of the system: such small net exports from vast areas made up the input needed by urban systems such as Vienna. This export was only about 2–3 per cent of the domestic extraction of energy in both rural cases. By contrast, according to the figures presented here, Vienna depended almost completely on imports, which were mostly in the form of biomass: coal constituted only about 3 per cent of Vienna's energy input in 1800/1830 (Table 5.2).

The amount of energy flowing into the system in 1800/1830 was much lower in Vienna than in all other systems. This discrepancy is mostly a result of the fact that the energy imported to Vienna was different from the energy extracted in the rural systems: while the domestic extraction in the rural systems was biomass containing a lot of feed for livestock required to produce food for domestic consumption and export, Vienna imported 'ready-to-use' biomass that could be directly used as food, feed for draught animals and fuelwood. Fuelwood and charcoal for space/water heating as well as manufacture accounted for about 80 per cent of Vienna's energy

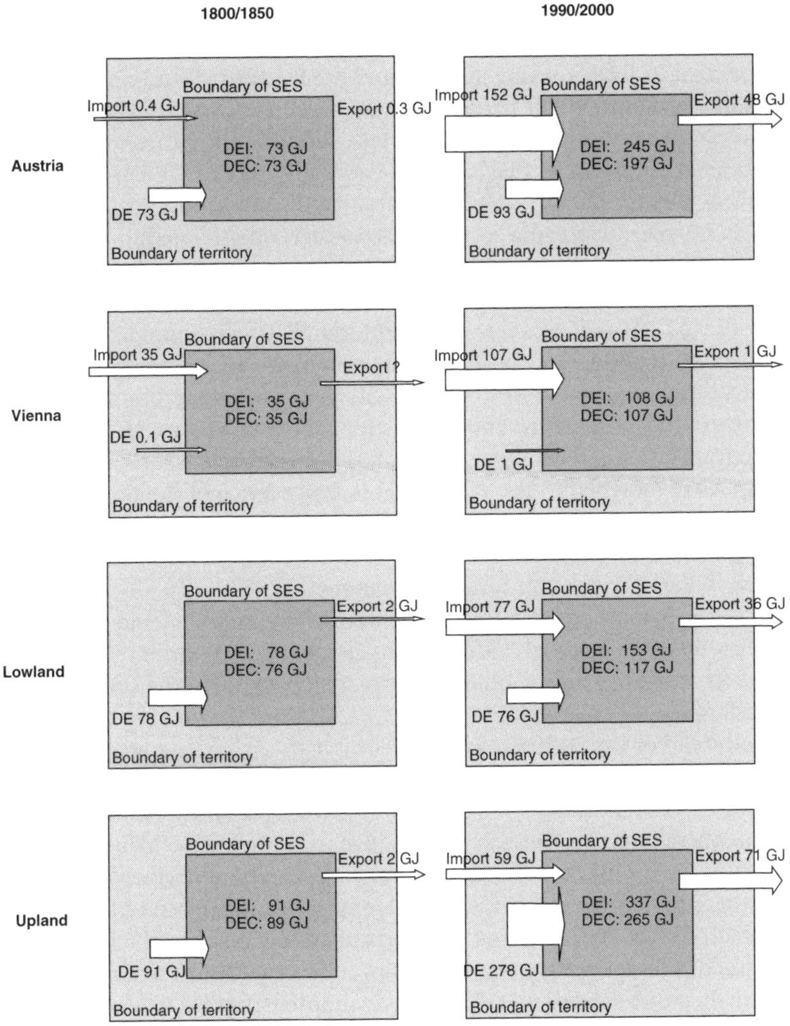

Notes: Values are gigajoules per capita and year (GJ/cap/yr). SES is an abbreviation for 'socioeconomic system' with the same system boundaries as applied on the national scale (see Chapter 2). DE denotes 'domestic extraction', DEI 'direct energy input' and DEC 'domestic energy consumption'.

Source: Table 5.2.

Figure 5.1 Main components of energy flow analyses for Austria, Vienna, the lowland and the upland systems

import, food for humans for about 17 per cent and 4 per cent was fed to draught animals (including about 5000 horses). That is, much less energy was needed to feed working animals and other livestock in Vienna than in the rural systems, which required about 40–50 GJ/cap/yr to sustain their livestock, more than half of their total energy input. Moreover, the per-capita consumption of fuelwood was actually lower in Vienna than the Austrian average, despite the fact that a significant amount of wood was used in the manufacturing sector. This was certainly related to the relative scarcity and hence also more efficient use of wood in the city.

Under the conditions of the agrarian regime, urban/industrial centres were fuelled entirely by their rural hinterland, which supplied the urban dwellers with food, firewood and raw materials. In energetic terms, the physical exchange processes between cities and the rural hinterland were unidirectional, cities represented a sink for energy and plant nutrients. Only comparatively small amounts of material (textiles, iron goods) but no energy were returned to the villages. This dependence of cities on biomass and, therefore, the energy surplus of the rural hinterland was a major bottleneck for urban growth and spatial concentration, an issue that is discussed in detail in Section 5.3 of this chapter.

In turn, urban/industrial centres increasingly supplied the hinterland with products with a high value density, for instance textiles or iron tools. This was, however, only a small material flow back from the centres to the hinterland.

By the end of the 20th century, Austria as a whole had become a through-put system in energetic terms, with the energetic value of imports being about three times as large as exports. Fossil fuels accounted for over 80 per cent of these imports. Rural systems had also become dependent on energy imports, as their inflows of fossil fuels and electricity were considerable. On the other hand, their ability to export biomass had increased tremendously, by a factor of 18 in the case of the lowland system and by a factor of 36 in the case of the upland system. It is noteworthy that the lowland system had already become dependent on net energy imports, with a relation between imports and exports of over 2, whereas the upland system still exported slightly more energy than it imported. This shows that rural–urban relations had changed their character fundamentally, as rural systems, thanks to the availability of fossil fuels, they no longer had to fulfil a role as suppliers of net energy to urban systems, a role they had inevitably had in the agrarian regime. Except for systems including fossil-fuel extraction sites, hydropower plants or other similar installations, only predominantly agricultural systems can today be expected to export more energy than they consume. The negative energy balance of the lowland system can be explained on the one hand by its function as a settlement area for people commuting to nearby centres, above all, to the Lower Austrian capital St Pölten, and on the other by the meanwhile significant presence of economic activities outside agriculture, that is, firms in the manufacturing and service sectors.

The changes seem to be smallest for Vienna: its energy throughput per capita had tripled, but it continued to depend completely on energy imports. The fuel mix had changed substantially, however: about 88 per cent of the imports at the end of the 20th century were fossil fuels, the remainder were electricity and biomass, mostly food. Vienna still consumed considerably less energy than the Austrian average, and the reason remains that Vienna mainly imported 'ready-to-use' final energy and most energy conversions required to produce this energy took place outside its boundaries. Moreover, Vienna was dominated by service sectors, a considerable share of the population being engaged in government, education, company headquarters and so on, and had little heavy industry. It may be assumed that Vienna imported large amounts of so-called 'grey energy', that is, energy incorporated in the products it consumed but that were produced elsewhere. To establish how large this amount of energy consumed only indirectly was would require an input–output analysis to be undertaken.

The livestock-intensive agricultural production system of the upland system is reflected in an extremely high per-capita value of domestic extraction: in the upland system the per-capita domestic extraction of biomass was larger than Austria's aggregate per-capita energy input, and about five times higher than the average Austrian per-capita domestic extraction of biomass.[4]

The extent of specialization between different rural systems, which was low in 1800/1830, has, by the end of the 20th century, become striking. In the early 19th century, the per-capita domestic extraction of Austria as a whole and of the lowland as well as the upland system was astonishingly similar. At the end of the 20th century, extreme differences are visible. This specialization process can be traced by looking at the indicators displayed in Table 5.1: in 1800/1830 livestock density was identical in the lowland and the upland systems – in both cases, 24 livestock units (LU) per km² – and similar to the Austrian average of 17 livestock units per km². Nowadays, livestock has almost disappeared from the lowland system (8 LU/km²), while the density has almost tripled in the upland system to 61 LU/km².

To summarize: we find that the most significant change between agrarian and industrial metabolism, besides the surge in overall throughput and the reliance on new resources such as fossil fuels, is the increasing connectedness and interdependence between different locales. In the agrarian regime, almost all material and energy flows were local, and beyond that there were small, but largely unidirectional flows from rural areas (sources) to urban areas (sinks). A similar flow existed also for population, with rural areas exporting people to urban centres, which, at least during the early stages of the transition, were not able to sustain themselves demographically (Wrigley 1985). The biomass 'exported' consisted partly of tithes and taxes and partly of produce sold on the market. Revenues were mostly used to buy essential goods such as iron tools, salt and other items that could not be produced locally.

In the industrial regime, within a highly developed country, energy flows between urban centres and rural communities become more symmetrical, as they flow in both directions. These exchange processes are fuelled by internationally organized flows of fossil fuels. In the case discussed here, the flows between urban and rural systems grew by a factor about one order of magnitude larger than the growth of the per-capita or national-level flows discussed in Chapter 2. For example, Austria's per-capita DEC grew only by a factor of 2.7, but the outflows from the rural systems grew by factors between 18 and 36. Economically, these often marginal rural systems depend to a large extent on subsidies making up a significant part of the income of agricultural households. Retired people account for a substantial share of the population in many marginal rural communities (among them the lowland system discussed here), their income is obtained from transfers (pensions), as is the income of unemployed people, proportionately more of whom live in rural than in urban regions. Moreover, in many rural communities a significant percentage, sometimes even the majority, of gainfully employed people commute to regional centres for work, which is also a strong factor in the lowland case discussed here.

5.3   SCALE LINKAGES AND THEIR IMPLICATIONS FOR TRANSPORT AND SPATIAL ORGANIZATION

The analyses presented in the previous section have shown that the transition from an agrarian to an industrial society implies enormous increases in the flows of materials and energy between different locales. That is, transport processes of materials as well as people are soaring. In this section we will discuss 1) implications for overall transport intensity and 2) their significance for land use.

Rural–Urban Interrelations and Transport Intensity

Using the data presented above, it is possible to estimate the area demand of urban centres. Vienna's energy import of about 35 GJ/cap/yr or about

8 petajoules per year (PJ/yr) in the early 19th century was equivalent to about 600 000 tons of biomass, above all, wood, grain, meat, milk, feed and so on. Assuming typical yields and conversion efficiencies of the early 19th century, the supply of this amount of materials and energy required about 70–100 times more area than was available within the boundary of the city: Vienna's wood and food demand each directly required about 1500–3000 km^2, and feed for draught animals required about 150 km^2. The supply of an average inhabitant of Vienna required about 1.5–2 hectares (ha) of agricultural or forest land, of which only about 0.02 ha were available within the boundaries of the city.

The area actually used to support Vienna's 230 000 inhabitants was, however, much larger: as shown above, only a small fraction of the production of rural areas could actually be exported to support urban centres. According to the data presented above (see also Fischer-Kowalski, Krausmann and Smetschka 2004), between five and 15 people living in rural communities were required to produce enough food, fuelwood and feed to support one city dweller. Therefore, the total amount of area indirectly needed per capita was probably about 10–30 hectares.

This illustrates the limitations for urban growth under the conditions of the agrarian regime: comparatively low yields and agrarian surplus rates translated urban demand into vast hinterlands required to supply the city with sufficient energy and raw materials. Together with the high energy costs of transport this constituted a major bottleneck for urban/industrial concentration. In particular, overland transport was associated with high costs that were prohibitive for long-distance transport of bulk materials such as firewood or cereals. Waterways provided a much cheaper means of transport and inland cities such as Vienna relied heavily upon rivers and canals for their supply. Hence, not the absolute amount but the location of woodlands in relation to waterways was decisive for urban wood supply. Only about 50 per cent of Vienna's wood demand could be supplied from adjacent forests, but large amounts of wood originated from distant woodlands along the Danube river: some 26 per cent came from Upper Austria, Salzburg and Tirol and about 10 per cent from Bavaria, all of which was transported several hundred kilometres on the Danube and its feeders (Wessely 1882). Grain came partly from regions that today form part of the Czech Republic or Hungary; two-thirds of the oxen consumed as meat came from Hungary and were walked across the Puszta, thereby creating a very specific grassland ecosystem that is all but 'natural' (Mayr 1940; Messing 1899).

In the early 20th century, Vienna's population had risen to 1.6 million. Despite considerable increases in agricultural yields, the direct area required (that is, not counting the support of the agricultural workforce) to provide the food consumed by this roughly five times larger population had risen to about 6000–10 000 km². Vienna's 40 000 draught animals required another 1000 km². The city's demand for combustible fuel had grown almost tenfold since 1800. Assuming that the 60 PJ of fuel burned annually in Vienna would have had to be supplied from fuelwood alone, this energy demand translates into a hypothetical forest hinterland of 20 000 km². This is a vast area, equivalent to about two-thirds of the Austrian woodlands at that time and it is easy to imagine that it would have hardly been possible to manage the fuel supply of Vienna on the basis of wood alone. By 1910, however, 95 per cent of the technical energy was supplied from coal, which increasingly substituted for fuelwood from the 1840s onwards. The actual forest areas needed to supply wood had shrunk to about 1000 km², only one-third of the area needed 100 years earlier. Taken together, these figures imply that Vienna's direct area requirement per capita sank to about 0.5–0.75 ha/cap/yr. Both yields and labour efficiency had risen considerably by the early 20th century so that the indirect area requirement per capita must have been reduced considerably too.

Due to the enormous increases in yields, the direct area requirement of Vienna's food supply is today in the same order of magnitude as it was in the early 19th century, that is, about 3000 km², despite a considerable rise in the share of meat and other animal-derived food, and despite the population being approximately five times larger. The indirect area requirement of agricultural workers is negligible, due to the more than 300-fold increase in agricultural labour productivity discussed above. Of course, all of this was only made possible by the use of fossil fuels. Today, substituting biomass-derived fuels for fossil fuels consumed in Vienna would require about 25 000–30 000 km², roughly one-third of Austria's territory (Krausmann 2005).

All of this implies that enormous increases in freight transport were a prerequisite for the emergence of modern consumption patterns. Unfortunately, at present no consistent time series for the transportation volumes required for the supply of Vienna is available. In a recent modelling exercise, Fischer-Kowalski et al. (2004) estimated that about 800 000 people were required to feed an urban centre of 100 000 inhabitants in the 19th century. Freight transport amounted to about 11 ton-kilometres per capita and year (tkm/cap/yr).[5]

Figure 5.2 presents data on the amount of transport required for Austria's biomass supply from 1926, showing that Austria's foreign biomass trade already required about 300 tkm/cap/yr in the 1930s, a number that surged to a staggering 2200 tkm/cap/yr in the year 2000. If we add to these data rough estimates of internal (on-farm) transport and domestic transport derived from the model calculations discussed above (Fischer-Kowalski et al. 2004) we arrive at the result presented in Table 5.3.

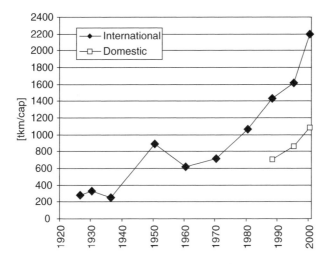

Sources: Unpublished calculations based on Erb (2004); Erb (pers. comm).

Figure 5.2   *International and domestic transport required for Austria's biomass supply, 1926–2000*

Table 5.3   *Transport demand of biomass supply in Austria, 1830 and 2000: a first estimate*

	1830 [tkm/cap/yr]	2000 [tkm/cap/yr]	Growth Factor
Local (on-farm) transport	4	7	1.8
Domestic commercial	7	1100	160
International commercial	n.d.[a]	2200	4
Total	11	3300	300

Note: [a] Near zero.

Sources: Fischer-Kowalski et al. (2004); Herry (2003); own estimates.

Even though we admit that these are first, rough estimates, we are confident that we have correctly estimated the orders of magnitude at the least. In any case, this analysis shows that changes in transport technology and infrastructure are among the most decisive forces driving transitions from agrarian to industrial society, and with them all their implications for land use, socioeconomic metabolism and other environmentally relevant aspects of society–nature interaction (see also Ciccantell and Bunker 1998; Grübler 1998).

Impacts on Land-use Patterns and Nutrient Flows

As the transition from the agrarian to the industrial regime resulted in spatial specialization between different regions, it had profound impacts on spatial patterns of agriculture in Austria on the national scale (Krausmann et al. 2003). Many of these changes can only be understood by analysing the role of livestock. In the agrarian system, livestock formed an indispensible part of agriculture, not only for food production but above all as working animals and for their function in the nutrient cycle.

Therefore, even in fertile lowlands a mix of cropland, grassland and forest prevailed to guarantee the supply of food, feed and draught power as well as wood for fuel and construction materials. On the other hand, even in mountainous regions, cropland was needed to produce plant food for humans (Krausmann 2001; Netting 1981; Project Group Environmental History 1999; Sieferle 1997; Winiwarter and Sonnlechner 2000).

Fossil fuels and other external energy sources resulted in various concentration processes: intensive cropland farming was concentrated in fertile lowlands and more or less abandoned in mountainous regions. Cattle rearing receded from the intensive cropland regions, which in turn led to a considerable reduction in grassland area there. Manure, formerly an essential source of plant nutrients, was replaced by mineral fertilizer. The use of mineral fertilizer also meant that less area had to be planted with leguminous crops, such as clover, that were used in crop rotation schemes to fix nitrogen and as cattle feed. The fattening of pigs, poultry and cattle for meat production, mostly based on fodder from cropland such as barley, maize and pulses, was concentrated in regions suitable for maize or fodder cereal cultivation but less competitive in wheat and rye production. Mixed agriculture – for example, Simmental cattle farming for combined milk and meat production – retreated to regions not suitable for large wheat and maize monocultures. Such forms of agriculture survived in fertile, hilly, pre-alpine regions and in the ancient granite stock in northern Austria (Böhmische Masse). In the high alpine regions, only grassland agriculture remained, dominated by cattle farming and some sheep rearing.[6]

One important aspect of these changes relates to the growing importance of long-distance transport processes within the agricultural sector, including their significance for nutrient flows. Using factors of species-specific fodder consumption (for example, Hohenecker 1981) and combining them with grassland as well as cropland yield data, we derived feed balances for Austria's municipalities in 1960 and 1995. Feed production and feed demand was still roughly balanced in many municipalities in 1959. Few municipalities existed where either feed demand was much higher than feed production or vice versa. The picture changed completely in 1995: large grain-producing regions emerged, producing considerably more feed than they consumed, for example, the fertile lowlands in the northeast of Austria. On the other hand, large, coherent 'feed deficit' regions emerged in the hilly, pre-alpine regions of Upper and Lower Austria in which cattle and pig densities are high (see Krausmann et al. 2003).

This implies that a large part of the animal feed used in contemporary Austria in 1995 had to be transported over considerable distances. Of course, this also means that previously rather closed, local nutrient cycles, such as that of nitrogen, now extended over large distances: nitrogen entered intensive cropland regions as mineral fertilizer and was then transported as feed to the feed-deficit regions, where nitrogen in manure, by far exceeding local requirements, was then discharged on grasslands. Nitrogen contained in animal products was then transported to urban centres, where it entered the sewage water treatment system and eventually ended up in sewage sludges that are, in Austria, mainly deposited. The once mainly cyclical flow of nitrogen has, thus, been turned into a largely unidirectional flow from air to factory to agro-ecosystem to humans to final repository.

What this means for rural regions is shown in Figure 5.3, which presents nitrogen flows expressed as kilograms of pure nitrogen per hectare of agricultural area per year (kgN/ha/yr) in the lowland system.

According to this assessment (Krausmann 2004), imports of nitrogen increased from practically nil to almost 70 kgN/ha/yr and nitrogen exports increased to 40 kgN/ha/yr – each of these two flows is by far larger than the aggregate nitrogen turnover in 1830. Nitrogen export increased by a factor of about 20 and nitrogen contained in harvested biomass by a factor of about 3.[7] As this figure shows, almost closed local systems were replaced by throughput systems during the transition from the agrarian to the industrial regime. A much larger-scale pattern has emerged, which can only be sustained through massive, continuous inputs of fossil fuels and a large-scale transport infrastructure.

5.4   CONCLUSIONS

The transition from the agrarian to the industrial socioecological regime implies not only the changes in resource use, land use and technology discussed in Chapter 2 but also results in a fundamental reorganization of spatial patterns in socioecological systems. These changes are highly relevant for social organization, including rural–urban relations, spatial patterns in the division of labour and many other aspects of socioeconomic systems. They are equally relevant for ecosystems and cultural landscapes,

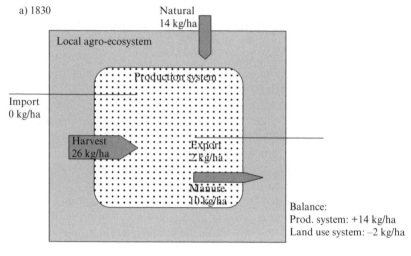

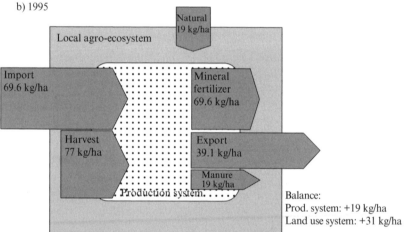

Source: Krausmann (2006b).

Figure 5.3   Yearly nitrogen flows in the lowland system in 1830 and 1995

having changed nutrient cycles, spatial patterns in the landscape, spatial organization of different kinds of land use, biodiversity and other factors.

From a socioeconomic point of view, one particularly striking aspect is the enormous growth in transport volumes, measured as ton-kilometres, associated with the spatial reorganization process discussed in this chapter. The 'growth engine' (Ayres and van den Bergh 2005) – that is, the positive feedback loop – behind this seems to be the growing division of labour, which implies longer chains of production, and the utilization of economies of scale in each of the steps of production. These two factors combined clearly mean that more intermediate products have to be transported over ever-longer distances, which is obviously only possible because transport costs are relatively low, due partly to the availability of publicly financed (in other words, subsidized) transport infrastructures.

What a possible substantial rise in transport costs, which might follow a decline in global oil production, could mean for this dynamics is relatively easy to guess, although the entire consequences are difficult to imagine. Global oil production would decline, should the world eventually reach the maximum of global oil production. This event is often termed 'peak oil' and is predicted by some (Campbell 1997, 2004; Hallock et al. 2004)[8] to be likely to occur in the next 15–20 years, perhaps even earlier. From the analysis presented here it is clear that such an event could render much of the current transport infrastructure unsustainable, and would thus call into question current settlement patterns, economic structures and trajectories, the organization of production processes, including agriculture, the distribution of goods and many other fundamental traits characteristic of the organization of modern socioeconomic systems.

From an ecological perspective, the analysis reveals the intricate interrelations between human energy systems, including energy resources used, dominant technologies or even technological clusters (Grübler 1998), and between energy prices (Fouquet and Pearson 1998, 2004) on the one hand and ecosystems on the other. In densely populated countries such as Austria, socioeconomic drivers strongly influence ecological patterns and processes such as primary production (Haberl et al. 2001b), spatial patterns in landscape ecosystems (Wrbka et al. 2004), biodiversity (Erb 2004; Haberl et al. 2004b; Haberl et al. 2005) and many others. How socioeconomic and natural forces interact over long periods of time in shaping cultural landscapes emerging in this interaction process, is an important research question (Haberl, Batterbury and Moran 2001a). Further multi-scale analyses, such as the one presented here, are required if the spatial dimensions of socioecological transitions are to be understood.

APPENDIX 5.1   METHODS USED TO ESTIMATE FOSSIL FUEL USE IN NUSSDORF AND KOTTES

Current fossil fuel energy use in Nussdorf and Kottes was estimated as follows. The estimation was mainly based on the 1991 census, which provided data on household numbers, dwelling size, fuels used for space and water heating, occupation by sector and so on. Final energy use in private house-holds was estimated considering three different kinds of end use: 1) space and water heating, 2) electric appliances including light, and 3) transportation. Final energy used for space and water heating was based on detailed calculations that reflected size and age classes of dwellings, a breakdown of buildings into single family houses and multiple dwelling units, as well as figures on heating systems and fuels used. These indicators were used together with factors on average efficiencies and consumption per unit in the Austrian average taken from the literature (Bertsch et al. 1995). Electricity consumption of electrical appliances was estimated using the average per-capita household electricity consumption in Lower Austria minus the estimated amount of electricity used for space and water heating that resulted from the previous calculation. Fuel use for cars was estimated using data on car stocks per 1000 inhabitants for the respective political districts (St Pölten Land and Krems), assuming an average fuel consumption of 36 GJ/cap/yr (Bertsch et al. 1995).

To calculate energy use in agriculture we used data of diesel fuel use per hectare of cropland, permanent cultures and forest (Haberl et al. 2002). These estimates were derived from official data on the costs of agricultural inputs per hectare and crop published by the Austrian Ministry of Agriculture (BMLF 1992). Using data on cropped area in Kottes and Nussdorf, these data could be used to calculate the use of diesel fuel. Diesel fuel use amounts to 60 per cent of total final energy use in Austria's agriculture (Alder 1999). Total agricultural final energy use was extrapolated from diesel fuel use, that is, we assumed that the distribution of fuels in the lowland and the upland systems is similar in Austria as a whole. For this extrapolation, we used data from an in-depth appraisal of energy use in agriculture and forestry produced by Statistics Austria (Alder 1999).

An estimate of final energy use in other sectors (manufacture, commercial/services and so on) was based on the number of workers. Numbers of workers were taken from the so-called workplace count (*Arbeitsstätten-zählung*), which is available with a breakdown into about 65 economic sectors. Final energy use per worker was determined by using Austria's official sectoral energy balance (Alder and Kvapil 1994). Because the sectoral energy balance distinguishes considerably fewer economic sectors (42) than the workplace count, some sectors had to be aggregated in order for these calculations to be performed.

ACKNOWLEDGEMENTS

This chapter builds on material from a number of empirical studies on the relation of land use and social metabolism in Austrian villages in the 19th and the 20th century. The research was funded under the research programme 'Austrian Landscape Research' (http://www.klf.at/) of the Austrian Ministry of Education, Science and Culture and by the Austrian Science Fund within the project 'The Transformation of Society's Natural Relations' (Project No. P16759).

Methodological aspects and detailed analysis of the data used in this chapter have been published in a number of books and journal articles, above all, Krausmann (2004, 2006a, 2006c); Projektgruppe Umweltgeschichte (1997, 1999); Sieferle et al. (2006); and Winiwarter and Sonnlechner (2001).

The authors are grateful to Verena Winiwarter, Christoph Sonnlechner, Ortrun Veichtlbauer and Klaus Ecker for providing us with empirical material and their cooperation and to Marina Fischer-Kowalski, Rolf Peter Sieferle, Enric Tello, Geoff Cunfer, Karl-Heinz Erb and Simone Gingrich for their useful comments at various stages of the project.

NOTES

1. Note that we refer here to a notion of energy throughput that includes not only technical energy – that is, energy transformed in machinery – but also nutritive energy consumed by humans and livestock. In other words, we are concerned with the whole 'energetic metabolism' of society, as analysed in energy flow accounts (EFA) described in more detail in Chapter 2 and the literature (Haberl 2001a; 2001b).
2. Flows not recorded by statistics were modelled: the domestic extraction of biomass was calculated based on data for agricultural land within the administrative boundary of the city and assumptions on typical yields. Imports of food and feed for draught animals were calculated by applying figures for per-capita consumption from statistical sources and secondary literature and data on population and on urban stocks of draught animals. Methods for estimating current fossil fuel use are discussed in Appendix 5.1 at the end of the chapter.
3. According to the statistical data, one single family house in each, the lowland and the upland systems, uses district heating. However, since district heating requires a grid, which is obviously present in neither of the two communities, we assumed that these were statistical flaws and that both houses are actually using heating oil.
4. Of course this depends on the form of standardization: as there are few people and many animals, values calculated per capita of humans will necessarily be high. A standardization by area would give different results.
5. For example, 1 ton could have been transported over a distance of 11 km, or 11 tons could have been transported over a distance of 1 km to arrive at this value.
6. These processes are analysed and mapped in detail in a recent paper (Krausmann et al. 2003), which, among others, presents maps of cattle and pig density in 1960 and 1995 as well as changes in the grassland/cropland ratio during this period.
7. Similar analyses of human-induced changes in nitrogen flows have, among others, been conducted by Robert Ayres and colleagues (Ayres, Schlesinger and Socolow 1994; Domene and Ayres 2002).
8. We note that this issue is controversial and we are by no means experts in this field. The likelihood or otherwise of peak oil happening in the next two decades is beyond the scope of our judgement. However, that oil reserves are finite, for practical purposes, and that the world therefore must eventually reach a maximum of oil production during the next couple of decades, is contested only by a few analysts (for example, Odell 2004).

REFERENCES

Alder, Robert (1999), 'Energieeinsatz in der Land- und Forstwirtschaft 1997', *Statistische Nachrichten*, **54**(7), 579–82.

Alder, Robert and Brigitte Kvapil (1994), 'Energieaufkommen und -verwendung in der österreichischen Volkswirtschaft im Jahre 1991, Endgültige Energiebilanz 1991', *Statistische Nachrichten*, **49**(8), 695–707.

Ayres, Robert U. and Jeroen C.J.M. van den Bergh (2005), 'A Theory of Economic Growth with Material/Energy Resources and Dematerialization: Interaction of Three Growth Mechanisms', *Ecological Economics*, **55**, 96–118.

Ayres, Robert U., William H. Schlesinger and Robert H. Socolow (1994), 'Human Impacts on the Carbon and Nitrogen Cycles', in Robert H. Socolow et al. (eds), *Industrial Ecology and Global Change*, Cambridge: Cambridge University Press, pp. 121–57.

Bertsch, Elmar, Johannes Fechner, Edith Zitz, Franz Schweitzer, Karl Lummerstorfer, Johannes Haas, Eckart Drössler, Harald Rohracher, Waltraud Winkler-Rieder, Helmut Haberl, Michael Bockhorni, Robert Thaler, Waltraud Frosch and Roman Riedel (1995), *Leitfaden Klimaschutz auf kommunaler Ebene*, Vienna: Bundesministerium für Umwelt.

BMLF (1992), *Bericht über die Lage der österreichischen Landwirtschaft 1991*, Vienna: Bundesministerium für Forst- und Landwirtschaft.

Boserup, Ester (1965), *The Conditions of Agricultural Growth. The Economics of Agrarian Change Under Population Pressure*, Chicago: Aldine/Earthscan.

Boserup, Ester (1981), *Population and Technological Change – A Study of Long-term Trends*, Chicago: The University of Chicago Press.

Campbell, Colin J. (1997), *The Coming Oil Crisis*, Essex: Multi-Science Publishing.

Campbell, Colin J. (2004), 'The Peak and Decline of World Oil Supply', *Energy*, 1/2004, 10–11.

Ciccantell, Paul S. and Stephen G. Bunker (1998), 'Introduction: Space, Transport, and World-Systems Theory', in Paul S. Ciccantell and Stephen G. Bunker (eds), *Space and Transport in the World-System*, Westport, CT and London: Greenwood Press, pp. 1–15.

Domene, Alejandra F.L. and Robert U. Ayres (2002), 'Nitrogen's Role in Industrial Systems', *Journal of Industrial Ecology*, **5**(1), 77–103.

Erb, Karl-Heinz (2004), 'Actual Land Demand of Austria 1926–2000: A Variation on Ecological Footprint Assessments', *Land Use Policy*, **21**(3), 247–59.

Fischer-Kowalski, Marina, Fridolin Krausmann and Barbara Smetschka (2004), 'Modelling Scenarios of Transport Across History from a Socio-metabolic Perspective', *Review. Fernand Braudel Center*, **27**(4), 307–42.

Fouquet, Roger and P.J.G. Pearson (1998), 'A Thousand Years of Energy Use in the United Kingdom', *The Energy Journal*, **19**(4), 1–41.

Fouquet, Roger and P.J.G. Pearson (2004), 'Seven Centuries of Energy Services: The Price and Use of Light in the United Kingdom (1300–2000)', *The Energy Journal*, **25** 1–34.

Grübler, Arnulf (1998), *Technology and Global Change*, Cambridge: Cambridge University Press.

Haberl, Helmut (2001a), 'The Energetic Metabolism of Societies, Part I: Accounting Concepts', *Journal of Industrial Ecology*, **5**(1), 11–33.

Haberl, Helmut (2001b), 'The Energetic Metabolism of Societies, Part II: Empirical Examples', *Journal of Industrial Ecology*, **5**(2), 71–88.

Haberl, Helmut, Simon P.J. Batterbury and Emilio F. Moran (eds) (2001a), *Using and Shaping the Land: A Long-term Perspective. Special Issue of Land Use Policy*, **18**(1), Oxford: Pergamon/Elsevier.

Haberl, Helmut, Fridolin Krausmann, Karl-Heinz Erb, Niels B. Schulz and Heidi Adensam (2002), 'Biomasseeinsatz und Landnutzung Österreich 1995–2020', Wien: IFF Social Ecology (Social Ecology Working Paper; 65).

Haberl, Helmut, Marina Fischer-Kowalski, Fridolin Krausmann, Helga Weisz and Verena Winiwarter (2004a), 'Progress Towards Sustainability? What the Conceptual Framework of Material and Energy Flow Accounting (MEFA) Can Offer', *Land Use Policy*, **21**(3), 199–213.

Haberl, Helmut, Karl-Heinz Erb, Fridolin Krausmann, Wolfgang Loibl, Niels B. Schulz and Helga Weisz (2001b), 'Changes in Ecosystem Processes Induced by Land Use: Human Appropriation of Net Primary Production and its Influence on Standing Crop in Austria', *Global Biogeochemical Cycles*, **15**(4), 929–42.

Haberl, Helmut, Christof Plutzar, Karl-Heinz Erb, Veronika Gaube, Martin Pollheimer and Niels B. Schulz (2005), 'Human Appropriation of Net Primary Production as Determinant of Avifauna Diversity in Austria', *Agriculture, Ecosystems & Environment*, **110**(3–4), 119–31.

Haberl, Helmut, Niels B. Schulz, Christoph Plutzar, Karl-Heinz Erb, Fridolin Krausmann, Wolfgang Loibl, Dietmar Moser, Norbert Sauberer, Helga Weisz, Harald G. Zechmeister and Peter Zulka (2004b), 'Human Appropriation of Net Primary Production and Species Diversity in Agricultural Landscapes', *Agriculture, Ecosystems & Environment*, **102**(2), 213–18.

Hallock, John L. Jr, Pradeep J. Tharakan, Charles A.S. Hall, Michael Jefferson and Wei Wu (2004), 'Forecasting the Limits to the Availability and Diversity of Global Conventional Oil Supply', *Energy*, **29**(11), 1673–96.

Herry, M. (2003), *Verkehr in Zahlen Österreich*, Wien: Federal Ministry for Transport, Innovation and Technology.

Hohenecker, Josef (1981), 'Entwicklungstendenzen bei der Futterversorgung Österreichs, dargestellt am Beispiel ausgewählter Jahre', *Die Bodenkultur. Austrian Journal of Agricultural Research*, **32**, 163–87.

IPCC (2001), *Climate Change 2001: The Scientific Basis*, Cambridge: Cambridge University Press.

Krausmann, Fridolin (2001), 'Land Use and Industrial Modernization: An Empirical Analysis of Human Influence on the Functioning of Ecosystems in Austria 1830–1995', *Land Use Policy*, **18**(1), 17–26.

Krausmann, Fridolin (2004), 'Milk, Manure and Muscular Power. Livestock and the Industrialization of Agriculture', *Human Ecology*, **32**(6), 735–73.

Krausmann, Fridolin (2005), 'Sonnenfinsternis? Das Energiesystem von Wien im 19. und 20. Jahrhundert', in Karl Brunner and Petra Schneider (eds), *Umwelt Stadt. Geschichte des Natur- und Lebensraumes Wien*, Vienna: Böhlau Verlag, pp. 140–50.

Krausmann, Fridolin (2006a), 'Land Use and Socio-economic Metabolism in Pre-industrial Agricultural Systems: Four 19th-Century Austrian Villages in Comparison', Vienna: IFF Social Ecology (Social Ecology Working Paper; 72).

Krausmann, Fridolin (2006b), 'Landnutzung und Energie in Österreich 1750 bis 2000', in Rolf P. Sieferle et al. (eds), *Das Ende der Fläche. Zum Sozialen Metabolismus der Industrialisierung*, Vienna: Böhlau.

Krausmann, Fridolin (2006c), 'The Transformation of Central European Land Use Systems: A Biophysical Perspective on Agricultural Modernization in Austria since 1830' (in Spanish), *Historia Agraria*.

Krausmann, Fridolin and Helmut Haberl (2002), 'The Process of Industrialization from the Perspective of Energetic Metabolism. Socioeconomic Energy Flows in Austria 1830–1995', *Ecological Economics*, **41**(2), 177–201.

Krausmann, Fridolin, Helmut Haberl, Niels B. Schulz, Karl-Heinz Erb, Ekkehard Darge and Veronika Gaube (2003), 'Land-use Change and Socio-economic Metabolism in Austria. Part I: Driving Forces of Land-use Change: 1950–1995', *Land Use Policy*, **20**(1), 1–20.

Mayr, Josef K. (1940), *Wien im Zeitalter Napoleons. Staatsfinanzen, Lebensverhältnisse, Beamte und Militär*, Vienna: Verlag des Vereins für Geschichte der Stadt Wien.

Messing, Ludwig (1899), *Die Wiener Fleischfrage mit Ausblicken auf Production, Gewerbe und Consumverhältnisse*, Vienna: Wilhelm Frick.

Netting, Robert M. (1981), *Balancing on an Alp. Ecological Change and Continuity in a Swiss Mountain Community*, London, New York, New Rochelle, Melbourne, Sydney: Cambridge University Press.

Netting, Robert M. (1993), *Smallholders, Householders. Farm Families and the Ecology of Intensive, Sustainable Agriculture*, Stanford, CA: Stanford University Press.

Odell, Peter R. (2004), *Why Carbon Fuels Will Dominate the 21st Century's Global Energy Economy*, Essex, UK: Multi-Science Publishing.

Peterseil, Johannes, Thomas Wrbka, Christoph Plutzar, Ingrid Schmitzberger, Andrea Kiss, Erich Szerencsits, Karl Reiter, Werner Schneider, Franz Suppan and Helmut Beissmann (2004), 'Evaluating the Ecological Sustainability of Austrian Agricultural Landscapes – The SINUS Approach', *Land Use Policy*, **21**(3), 307–20.

Project Group Environmental History (1999), 'Landscape and History: A Multidisciplinary Approach', *Collegium Anthropologicum*, **23**(2), 379–96.

Projektgruppe Umweltgeschichte (1997), *Historische und ökologische Prozesse in einer Kulturlandschaft*, CD-ROM, Bundesministerium für Wissenschaft und Verkehr.

Projektgruppe Umweltgeschichte (1999), *Kulturlandschaftsforschung: Historische Entwicklung von Wechselwirkungen zwischen Gesellschaft und Natur*, Vienna: CD-ROM, Bundesministerium für Wissenschaft und Verkehr.

Sieferle, Rolf P. (1997), *Rückblick auf die Natur: Eine Geschichte des Menschen und seiner Umwelt*, Munich: Luchterhand.

Sieferle, Rolf P., Fridolin Krausmann, Heinz Schandl and Verena Winiwarter (2006), *Das Ende der Fläche. Zum Sozialen Metabolismus der Industrialisierung*, Vienna: Böhlau.

Wessely, Joseph (1882), *Forstliches Jahrbuch für Oesterreich – Ungarn. Oesterreichs Donauländer. II. Theil: Spezial-Gemälde der Donauländer*, Vienna: Carl Fromme.

Winiwarter, Verena and Christoph Sonnlechner (2000), *Modellorientierte Rekonstruktion vorindustrieller Landwirtschaft*, Stuttgart: Breuninger-Stiftung, Schriftenreihe *Der Europäische Sonderweg*, Band 2.

Winiwarter, Verena and Christoph Sonnlechner (2001), *Der soziale Metabolismus der vorindustriellen Landwirtschaft in Europa*, Stuttgart: Breuninger Stiftung.

Wrbka, Thomas, Karl-Heinz Erb, Niels B. Schulz, Johannes Peterseil, C. Hahn and Helmut Haberl (2004), 'Linking Pattern and Processes in Cultural Landscapes. An Empirical Study Based on Spatially Explicit Indicators', *Land Use Policy*, **21**(3), 289–306.

Wrigley, Edward A. (1985), 'Urban Growth and Agricultural Change: England and the Continent in the Early Modern Period', *Journal of Interdisciplinary History*, **XV**(4), 683–728.

6. The local base of transitions in developing countries

Clemens M. Grünbühel, Simron J. Singh and Marina Fischer-Kowalski

6.1 INTRODUCTION

In this chapter, we look at the dynamics of rural communities in order to understand the local baseline of sociometabolic transitions. Our aim is to analyse how these communities organize their biophysical flows with their environment where an industrial transformation occurring at higher system levels produces changing framework conditions. While regional studies provide us with relevant quantitative time series evidence on the region's shift from one socioecological regime to another and the rate at which this occurs, local studies allow only snapshots of the socioecological profile at the base of these national economies in transition.

The three sites we focus on – Nalang in Laos, SangSaeng in Thailand and Trinket, an island in the Nicobar archipelago belonging to India – lie within the region of Southeast Asia. They differ in their history, political ecology, environmental setting, national context, ethnic composition, and – most importantly for this study – their economic portfolio and resource-use patterns. All three sites may be considered fairly 'remote' in terms of accessibility and 'backward' in relation to the overall economic development of their country.

The Northeastern Thai and the Central Lao people are part of the Thai-Lao cultural area (Wongthes and Sujit 1989), while the Nicobarese belong to the Austro-Asiatic group of insular Southeast Asia. While the Northeastern Thai practise permanent rice cultivation and the Lao practise a mixture of permanent rice farming and shifting cultivation, the Nicobarese engage primarily in fishing, hunting and horticulture, particularly in the form of coconut plantations for subsistence as well as for cash. Each of the three cases differ in resource-use profiles. The agricultural communities (Laos, Thailand) to some extent use domestic animals as labour resources, while for the Nicobarese, the most important (semi-) domesticated animals are pigs, which have a high sociocultural and ritualistic status but otherwise constitute a resource drain.

All three communities rely largely on a subsistence economy, that is, on domestically extracted resources, either through socially managed agriculture and forestry or by directly extracting resources from the wild. Yet, the communities are not isolated: social, economic, administrative and cultural interaction on multiple scales are part of their daily life experiences. For the two communities in mainland Southeast Asia, rice is a staple crop that grants a certain degree of food security. However, as with most subsistence systems, in addition to farming, the communities maintain livestock, hunt and gather from the immediate natural environment and some household members engage in paid off-farm labour in order to raise cash needed to pay for consumer goods, agricultural inputs (fertilizers, tools), taxes and investments.

The subsistence system in the Nicobars is substantially different. Their main staple today – rice – is obtained from the market. In order to be able to buy rice, the Nicobarese produce copra (dehydrated coconut flesh used for the extraction of coconut oil) and sell it to private traders or to their own cooperatives.[1] As additional food, fish extracted from the lagoon provide necessary proteins. Lately, their dependency on trade and the regional copra market has become problematic. Copra prices have fluctuated heavily due to an increasing regional trend to substitute coconut oil for cheap palm kernel oil. Future developments (for example, changes in Indian legislation, further world market turbulences of rice and/or copra) may force the Nicobarese to adapt to a changing political, social or natural environment.[2]

The three cases were investigated between 1998 and 2003 using a common methodology. To assess the biophysical parameters of the three communities, we adapted the accounting framework usually applied for national studies, namely material and energy flow accounting (MEFA) (Krausmann et al. 2004; Schandl et al. 2002) and in another adapted version, for the historical case studies reported on in Chapter 5 of this volume. However, in contrast to national-level studies that rely heavily on official statistics, data for the local studies had to be gathered by undertaking months of empirical fieldwork. In doing so, we followed common conceptual and methodological guidelines while devising novel on-site data collection and estimation methods.[3]

6.2   INTRODUCING THE THREE COMMUNITIES

Ban Nalang, Laos

Ban Nalang lies in the District of Fuang, Vientiane Province, about 200 km by car from the capital city Vientiane (Viang Chan). Fuang is ethnically mixed, as are most areas in Laos, with a majority of ethnic Lao (52 per cent), followed by the Kh'mu (17 per cent) (Durrenberger 2001; Graeber 2002). With an area of 1630 hectares, Nalang has a total population of 702 people, according to the 2001 census. The population density is 43 persons per km² and the average family size is 6.7 persons per household. The Nalang area is characterized by a low valley with adjacent mountain ranges to the west and east. Near the village is a limestone rock formation with a summit reaching approximately 700 m. The main landcover types include 'evergreen forest' in the lowlands, 'high density mixed deciduous forest' in the sloped areas, and 'limestone rock' for the Nalang landmark, the Phalang. The latter is not bare rock but includes a vegetation type consisting of plants such as ficus, shrubs and epiphytes. The precipitation for the entire area is approximately 1500–2000 mm per year.

Historically, the area has experienced a number of immigration waves, particularly during and in the aftermath of the Indochina War (1964–75). An unknown number of Phuan immigrated during 1988–89 from the district of Xiang Khuang, which was heavily affected by the war. The Phuan in Nalang have easily merged with the original group of Lao, although they are in frequent contact with their homeland and sometimes even intermarry with the Phuan in Xiang Khuang. The immigration of Kh'mu (Lao Theung) was a gradual process, although first arrivals came during the late 1980s. Most Kh'mu in Nalang originally come from an area in Vang Viang District, about 150 km from Fuang. These were originally swiddeners who were forced to move due to the degradation and scarcity of land in their area, which was earmarked for agricultural development by the government. An elected headman leads the village administratively, assisted by two deputies, one of whom is Kh'mu. All Kh'mu reside in the eastern sector of the village. All landless households and those without paddy fields in Nalang are ethnic Kh'mu. Although Nalang villagers accepted the immigrants, they were not able to endow them with sufficient land.

Nalang is characterized by a subsistence economy dominated by rice farming, mainly paddies, although a small amount of shifting cultivation is carried on as well. Nalang is therefore no different from the rest of the Lao cultural area, where the production of rice is the primary activity of farmers. Furthermore, like most Lao, the people of Nalang produce mainly glutinous rice (*Oryza glutinosa*), which is used for daily consumption. In exceptionally productive years, the harvest surplus is sold to other local villagers and to traders. In Nalang, rice is cultivated in a single season (rainy season). Cultivation relies on yearly precipitation and neither chemical fertilizers nor artificial irrigation are used. Paddies are invariably located at the bottoms of valleys, which are fed by a network of free-flowing irrigation channels, while the higher reaches are used for shifting cultivation, food gathering activities and as pasture for cattle.

Next to rice, bamboo shoots form the most important source of nutrition and are gathered along with a large variety of plants, roots, tubers, palms, insects and mushrooms. Hence, gathering of non-timber forest produce (NTFP) remains an everyday activity of the inhabitants of Nalang. Gathering, hunting and fishing deliver an important fraction of the protein intake. All waterways and ponds are fished by a variety of techniques from hooks to casting nets. The hunting of forest animals is widely practised in order to provide supplementary food. Villagers use traditional hunting utensils, such as traps, crossbows and home-made guns. Species hunted include rats, wild boar, squirrel, pangolin, wild goat, deer, mouse deer and snakes.

Through increasing contact with the outside world, the market and government agricultural policies, Nalang has adapted its economic portfolio over the last quarter of a century. In 1980, a road was constructed from the district capital Muang Fuang to the border town of Xanakham. The road was initially built by logging companies in order to facilitate large-scale extraction of timber. Following road construction, many of the valuable trees in the Nalang area were extracted and forest degradation intensified. Also, as a consequence of road construction, the village was relocated from the original site on the banks of the Namlang stream to the roadside. The road is poorly maintained, unsealed and is the cause of environmental disturbance in a variety of forms, such as dust and danger from timber-hauling trucks.

Consequently, the inhabitants of Nalang were faced with several external influences that served to integrate their subsistent economy with the market to some degree. For example, in the late 1990s, cucumber production was introduced in Nalang as an important cash crop during the dry season. After the harvest on the rice fields, cucumber gardens are created along the waterways on the area of the harvested fields. Cucumbers are cultivated solely for sale on nearby local markets. Farmers invest small amounts of fertilizer and pesticides and large effort in manual labour to produce an adequate product for the market. In addition, bananas are grown in small plantations and also retailed on local markets. Secondary to agricultural activities, the village economy relies on natural resource extraction from the forest. Timber is logged for use by the village as well as for sale to traders.

In the past, buffalo rearing had been another important feature of the Nalang agricultural system. The buffaloes had been used for ploughing fields and for transport purposes. In addition, they had high cultural significance, since the meat was regarded as a ceremonial food during rituals and feasts. It was also the main sign of wealth and was widely used as the currency for dowries. The influx of agricultural machinery, especially the motor-plough, since the mid-1990s has diminished the importance of the buffalo. Cattle are increasingly preferred for the purpose of sale, since maturing times are more rapid and maintenance easier.

Although more than 80 per cent of the village area remains covered by forest, some portions of the forest cover have experienced serious degradation in the past. Wide stretches of bamboo have replaced areas of high-density mixed deciduous forest. These areas were used for shifting cultivation and livestock grazing, and bamboo invaded during the first few years of succession. Today, large parts of the forest are regulated and shifting cultivation is no longer widely practised. Shrubs or low-density deciduous forest are a result of livestock grazing. While in the dry season buffaloes and cattle are left to graze the harvested rice fields, they are brought to the near reaches of the forest during the planting period and thus retain a state of open forest. A great deal of NTFP is gathered from these areas and access for the villagers is far easier than it is to the high-density forest.

Recent developments also require adaptation to pressure from external influences, such as government authorities and the growing needs of the inhabitants. Still, subsistence production remains the most important economic activity and has ensured that Nalang has little reliance upon processed and marketed foods. Most of the infrastructure and housing is created without dependence on foreign goods. The reliance on local resources remains high.

SangSaeng, Thailand

The village of SangSaeng is located in Northeast Thailand, in the south-eastern part of the province of Isan close to the borders of Laos and Cambodia. The village, with an area of 184 hectares (ha), has a population of 171 (1998), the population density thus being 93 cap/km^2. As a part of the Thai–Lao cultural and linguistic area, the Isan are ethnic Lao (similar to the majority population of the People's Democratic Republic of Laos). Despite the process of national integration and the significant influence of mainstream Thai mass media, the Isan speak Lao and maintain a series of cultural traits that distinguish them from the Central Thai. However, like most other Thai citizens, the Isan are Buddhists and farm rice. Several aspects of Isan culture are unique in the cultural mosaic of Southeast Asia. SangSaeng is characterized by sandy soils and was formerly covered by deciduous tropical rainforest with a five-month growing period, of which only very little remains. Yearly precipitation amounts to around 1400 mm.

As in the entire Isan plateau, the area is flat and there are no continuous natural waterways in the area. The only major waterway of Isan, the Mun river, passes some 20 km away from the community.

Rice is the primary agricultural product in Isan and provides the basis of the traditional subsistence economy. Together with Laos, Northern Thailand, Northwest Vietnam and the Chinese province of Yunnan, Isan forms the glutinous rice belt, which is unique in the world (Sakamoto 1996; Wongthes and Sujit 1989). Glutinous rice is the most prominent component of the Isan diet and is normally consumed at least three times a day. Distinct from Central Thailand, Isan features as a strong peripheral economy. It is without doubt the most disadvantaged region in Thailand. Ignored by modern tourism, characterized by a high level of deforestation (Ramitanondh 1989) and little agricultural mechanization, its inhabitants barely manage to subsist. Additionally, the Isan have to manage with infertile farmland, inadequate soil and numerous droughts (KKU-Ford Cropping Systems Project 1982). In this difficult setting, Isan farmers need to choose a set of strategies that allows them to cope with their natural environment and at the same time, to keep pace with the rapidly changing modern world.

The economy of the village of SangSaeng can be seen as resting upon three main pillars. The first is agriculture, forming the basis of material existence for the rural population. The people of SangSaeng produce rice mainly for their own consumption but also sell a portion of the harvest on the market. Cash crop rice is usually of the state-promoted varieties (*Oryza sativa ssp. indica*, non-glutinous), which the producer families hardly consume themselves. Only 5 per cent of 'Jasmine' rice produced in SangSaeng is consumed within the village. In comparison, 90 per cent of glutinous rice production is consumed by the producers themselves within a year. Apart from rice, SangSaeng has little other commercial production. Villagers undertake gardening and keep chicken and ducks solely to provide food for their own consumption. Buffaloes are bred as working animals, meaning that only cattle are destined for the local market.[4]

Hunting and gathering represent the second pillar of SangSaeng's local economy. This may be perceived as an ancient and outdated method of subsistence, but it is, in this case, a highly developed strategy that is well adapted to the local environment (Fukui 1993). Hunting and gathering helps to diversify the diet, especially during the dry season, when gardening is limited or made virtually impossible by the dry environment of the Northeast Thai plateau. A wide variety of animals, most prominently fish, birds, amphibians and insects, as well as different kinds of leaves, flowers, herbs, mushrooms and roots are extracted from the fields and remaining forests.

Recently, hired labour has become an important form of activity in SangSaeng. Thus, paid work forms the third pillar of SangSaeng's economy. Working outside the village in nearby cities, in Bangkok, or in the coffee and rubber plantations of Southern Thailand is common for most SangSaeng families. Paid temporary or seasonal work ranges from construction or factory work to household work, tourism or plantation harvesting. However, the labour migration is seldom permanent and workers return to their village at the peak of labour-intensive times (during transplanting and harvesting of rice). Yet labour migration fulfils an important function for the local economy and should not be seen as separate from village life. Migrants supply their families in the village with necessary money, receiving material goods in return. Since most jobs are found in construction and plantation harvesting, which are low-paid seasonal jobs (80–100 baht per day),[5] the supplies taken from home are essential to the migrant workers. In particular, the amount of rice required for consumption by a migrant worker is taken directly from the village to the location of work.

Peak working periods in SangSaeng's rice production occur during the preparation of rice nurseries, ploughing, transplanting and harvesting. All able-bodied family members are expected to work together. The highly seasonal nature of labour migration is due to city workers going back to their villages to help in the fields. Individuals not suited to performing heavy fieldwork, such as the elderly and children, are expected to perform minor household tasks, such as preparing food for the workers or preventing livestock from entering rice fields.

Trinket Island, Nicobar Archipelago, India

Trinket is one of the 24 islands that constitute the Nicobar archipelago in the Bay of Bengal. Located some 1200 km off the east coast of India, the Nicobar Islands are part of the larger Andaman and Nicobar Island archipelago, forming an 840-km-long arched chain in the ocean. Separated from the Andamans by miles of deep sea commonly known as the Ten Degree Channel, the inhabitants of the Nicobars had little contact with those of the Andamans before their colonial history. Most geologists think that these islands are the peaks of a submerged mountain range extending from Arakan Yoma (Myanmar) in the north, to Sumatra (Indonesia) in the south (Dagar, Mongia and Bandopadhyay 1991; Saldhana 1989; Sankaran 1998). As in many other situations within insular Southeast Asia, the rise of sea levels following the Pleistocene period led to the isolation of a once continental mass, resulting in its floral and faunal endemicity (Andrews and Sankaran 2002).

The Nicobar Islands, home to an outstanding tropical biodiversity, have a total land area of 1841 km² characterized by a variety of habitats such as mangroves, coral reefs, sandy beaches, dense forests, grasslands and wetlands (Andrews and Sankaran 2002). The landscape of the different islands in the archipelago varies greatly, ranging from flatlands to hilly areas to undulating meadows. The islands of Car Nicobar, Chowra and Trinket are flat. In contrast, Teressa, Bompoka, Kamorta, Katchal, Nancowry and Great Nicobar are rather hilly. The highest point in the Nicobars is Mount Thullier (642 m) on Great Nicobar, followed by Empress Peak (439 m) on Kamorta. Tillangchong and Kondul have undulating landscapes with elevations up to 300 m. Great Nicobar alone has large rivers, all of which are perennial. Most islands are surrounded by coral reefs that prevent easy accessibility.

Of the 24 islands, only 12 are inhabited largely by an indigenous population, the Nicobarese. The Nicobarese, roughly 30 000 in number (2001 census),[6] belong to the Austroasiatic family having migrated from the Malay-Burma coast some 2000 years ago. The inhabitants of Trinket, like their counterparts on the other islands, live in villages along the coast protected by natural bays and mangroves. Of the total 36.26 km² land area of Trinket, nearly half is dense tropical forest and a third is grassland. Mangroves cover roughly 3 km², mainly on the western stretches of the island where villages are located. The total population of Trinket in 2001 was 399 inhabitants distributed in 43 households, thus giving a population density of 11 inhabitants per km².

The inhabitants of Trinket, like the other Nicobarese, are engaged in a subsistence economy that combines hunting and gathering, fishing and rearing pigs with horticultural activities, namely raising coconut plantations. Of the total coconut harvest, half is processed into copra,[7] a third is fed to pigs, and the remainder is consumed by chickens and humans. Rice is obtained in exchange for copra (or cash) and is their main source of carbohydrate. The major source of protein for the Nicobarese is obtained from the sea in the form of fish, oyster-shells, sea-cucumber and other marine food varieties. Besides cultivating coconuts, some islanders maintain fruit and vegetable gardens and nearly all go hunting and gathering in the forest.

To a great extent, the Nicobar economy depends on hunting, gathering and fishing. This is combined with the production of copra that is destined for the market in order to allow them to buy or exchange desired products otherwise not available on the islands. Such an economic portfolio owes much to the islands' geographical location on a historically important sea route that connected the Indian sub-continent (and later Europe) to Southeast Asia. Hence, the islands have had contact with the outside world for at least 2000 years, if not more. The Nancowry harbour (one of the best natural harbours in the world) was constantly used during bad weather by passing ships. Ships that needed repair from damage incurred during their voyages also called upon these islands and used them as a stopping place. In the long and arduous sea voyages of early times that were highly dependent upon winds, the islands provided respite to the sailors who anchored off their coasts to replenish stocks of fresh food and water. In exchange for food (yams, coconuts, chickens, pigs, bananas etc.), the natives received iron and later cloth as well from these ships. Over time, the occasional barter trade became an integral part of the economic activities of the natives, even though this was of low commercial importance to the merchants themselves. Indeed, iron and cloth came to be regarded as precious and were regarded as symbols of wealth to be displayed at rituals and festivities.

From the middle of the 18th century, the islands were under the intermittent and weak control of the Danes. The second half of the 18th century witnessed growing British domination in terms of both the oceanic trade of the region and territorial power in India. The Nicobar Islands were officially taken over by the British from the Danes in 1869 for the purpose of controlling piracy in the central Nicobar Islands and 'to avoid the risk of such inconvenience as would be caused by the possible establishment of a foreign naval station in such proximity' (Temple 1903). It was in the interest of the British Government to foster trade in these islands. A system of general administration was set up that regulated trade through trading licences and royalty was charged on the volume of export. Under British rule, trade in coconuts became more pronounced. Valued for their cheap exchange rate, the Nicobar Islands began to be visited on a regular basis by vessels from Burma and India to obtain cargoes of coconuts. While coconuts formed the bulk of the export, the cargo also included varied quantities of arecanuts, empty coconut shells (used in the manufacture of hookah pipes for smoking tobacco), seashells, rattan, tortoise shells, sea-cucumbers, swiftlet nests and ambergris, the last three regarded as precious on the international market.

The success of British policy became evident as the number of vessels visiting the harbour increased steadily from 15 to 20 vessels in the 1860s to between 40 and 50 vessels in the 1880s (Man 1903, p. 192). The Nicobars were now no longer a mere source of food and a place of refuge but a place where it was profitable to trade in coconuts and other products. A regular system of trade with the Nicobarese became established and the Nicobarese in turn began to place great reliance on the imported commodities for their daily sustenance. Imports comprised an assortment of goods including silver, several varieties of cloth, rice, tools, tobacco and matches. Imported rice became a staple and replaced pandanus (a wild fruit rich in pulp and fibre) as the main source of carbohydrate, the former being easier to obtain and prepare compared with the tedious task of processing pandanus into an edible form. E.H. Man, a British colonial officer, in his report following the closure of the Penal Settlement at Nancowry in 1888, wrote: 'the Nicobarese, now fully appreciating the advantages of intercourse with the outer world, are more anxious than ever to encourage the visits of these traders, from whom they hope to procure supplies of such articles as they have long since obtained at Nancowry and learnt to regard almost as necessaries of life' (Temple 1903, p. 192).

In 1882, for the first time, the Nicobar Trading Company Limited based at Penang applied and obtained permission to reside and trade at Nancowry. One of the reasons why it was necessary for traders to take up residence on the islands concerned copra production. In the 1860s, Theodore Weber, the manager of the Samoa branch of the German company Godeffray und Sohn, pioneered a major invention that revolutionized commercial planting in the tropics. This was the discovery that coconut meat scooped from the shell and dried (copra) could be exported instead of coconut oil. On arrival, the oil could be extracted in Europe for use in many industrial products, with the residue served as valuable cattle feed. Weber's innovation laid the foundations for the copra industry (Meleisea 1980, pp. 1–3). It was not long before copra production was introduced in the Nicobars as well.[8] Traders bartered the coconuts from the Nicobarese and sun-dried them into copra, which was eventually exported. The tropical climate of the Nicobars allowed copra production only during the dry period of the northeast winds when sunshine was abundant. At first, copra comprised only a small part of the exported cargo in relation to coconuts. Over time, this pattern changed and most of the coconuts bartered in the Nicobars began to be converted to copra. Copra production required traders to take up residence on the islands for some part of the year. Furthermore, the Nicobars were gaining prominence as an important source for coconuts, which induced the traders to establish semi-permanent bases on the islands.

Nearly three-quarters of a century later, some of the more enterprising Nicobarese began to make their own copra and export it to their advantage and to the resentment of the traders who had earlier held a monopoly (*Annual Report* 1837–38). After World War II, in 1945, with the reoccupation of the islands by the British (from the Japanese), a West Indian company by the name of M/s. Akoojee Jadwet and Co. acquired the sole trading licence on the Nicobars following an agreement with the British. By this time, the industrial demand for copra had increased considerably, and the company insisted on buying copra instead of coconuts from the Nicobarese. At the same time, this company replaced barter exchange with a system of money and weights and measures to conduct business transactions. The

Nicobarese, now so dependent on trade, had no choice but to give in to the whims of the market. This marked the beginning of copra production as a central economic activity throughout the Nicobars.

6.3 BIOPHYSICAL INDICATORS OF TRANSITION

We can ask whether the three very different cases presented above can be ordered sequentially according to a theoretical pattern of transition. As theoretical guidance we use on the one hand Boserup (1981) and on the other, Sieferle and Weisz (Sieferle 1997, p. 192 and Weisz et al. 1999). Boserup distinguishes stages of agricultural intensification that follow increasing population density. Sieferle and Weisz et al. distinguish socio-ecological regimes by their dominant mode of energy utilization. Table 6.1 seeks to combine these two theoretical approaches into an overarching scale. The first column lists various production activities as practised by rural communities, ordered according to the intensity of colonization of the territory and the amount of yield extracted from a given area. The other columns distinguish stages of transition from a hunting and gathering regime (Stage 1), through subsistence farming (Stages 2 and 3), to increasing market integration (Stages 4 and 5), to finally a full-blown fossil-fuel-based industrial regime (Stage 6). The claim made is that there is an in-built irreversible dynamics from one stage to the next, driven by growing population density (the Boserup hypothesis) and, in its later stages, by the first indirect (Stage 5) and then direct (Stage 6) use of energy sources beyond land-bound solar energy (typically fossil fuels).

Using this scheme, we place the three cases in the stage they represent. In the case of Nalang, the pattern seems to be rather straightforward with a production profile combining hunting and gathering, swidden agriculture and subsistence farming. While clearly having an agrarian mode of production, the inhabitants of Nalang still hunt and gather to supplement their diet. This owes a great deal to the fact that they are embedded in a terrain that is suited to forests but not to agriculture. Furthermore, much of the agriculture in Nalang is for subsistence, and a small surplus of what the inhabitants produce for themselves is sold to the market. In this sense, we can say that Nalang corresponds to Stage 4.

SangSaeng seems to be further advanced into an agrarian mode of production and more integrated into a market economy. SangSaeng's inhabitants, besides producing a variety of rice that they consume themselves, also engage in the production of a special variety of rice for the market. Much of the land around them is utilized for agriculture, and the little that remains is degraded forest that provides them at best with a small amount

Table 6.1 *Schematic representation of transitional stages of locally dominant socioecological regimes*

Land Use and Production Practices	Stage 1	Stage 2	Stage 3	Stage 4	Stage 5	Stage 6	Deviant Cases
Foraging	+	+	+	+	(+)	(+)	+
Swidden and/or transhumance		+	+	+	(+)	(+)	
Permanent subsistence farming			+	+	+	+	
Sale of agriculture surplus				+	+	+	
Production for the market					+	+	+
Fossil-fuel-based industrial agriculture							
Position of case studies				Nalang	SangSaeng		Trinket

150

of food from hunting and gathering. Accordingly, it seems appropriate to place SangSaeng at Stage 5.

The Trinket case does not conform to the regular patterns suggested in Table 6.1. A large proportion of subsistence is achieved by hunting and gathering. Being an island, the population has the advantage of being able to go beyond its terrestrial space and obtain a large part of its dietary food from the sea. At the same time, Trinket Island, owing to its location on a historic sea route, established a strong trade network via the export of copra (until 100 years ago, coconuts), for which coconuts have to be harvested. The establishment of coconut plantations is an atypical horticultural activity: it does not have an annual cycle and the trees, once planted, fruit for about 100 years. All this requires fairly little labour. Coconut trees are not planted especially for the market and much of the produce is consumed locally, either as feed for pigs and chickens or by humans. Only the processing of coconuts into copra (for which about half of the coconut harvest is used) can be said to be an extra activity engaged in for the market. Nonetheless, the export of copra is an important economic activity by which the Nicobarese procure essentials, including their staple food, rice. In this sense, Trinket Island reveals a curious profile that combines hunting and gathering (fishing) with production of copra for the market. The inhabitants do not engage in any other form of agriculture, swidden or permanent. From time to time they do engage in cultivating horticultural gardens with a variety of tropical fruits, but only during festivals and ceremonies. Thus, Trinket would be classified as a deviant case that combines Stage 1 and Stage 5, skipping Stages 2–4.

From an energy-use point of view (Sieferle 1997), these stages can be classified into three or four regimes. The first would comprise the hunter-gatherers (Stage 1) who 'passively' use solar energy in the form of natural production of biomass within the territory they have access to (including the coastal areas and lagoons), and utilize what they can gather or catch. With this regime, they remain dependent upon the resource density their environment provides. The second regime would be agrarian, characterized by 'active' solar energy utilization, intervening in the natural environmental processes by agriculture and stock breeding and maintaining an altered state of their environment with human and animal labour. Within this regime, the resource density of the territory is manipulated to yield many more products for human use (and less other products) than would be the case in the absence of continuous human intervention. Still, there is an upper limit to improvement and to the yields that may be achieved.[9] This limit can only be overcome by the additional use of a land-independent source of energy – and here we enter the third socioecological regime. This additional energy use need not necessarily be apparent in local agriculture.

It may occur on the national level, for example, leading to the creation of a transport system that allows agricultural produce from remote areas to be utilized more conveniently and at lower cost than by human or animal carriers or animal-drawn carriages.[10] So on the local level we may see an intermediate stage between an agrarian (solar) and industrial (fossil-fuel-based) regime in the form of indirect fossil fuel utilization as exemplified particularly in Stage 5: large-scale production for the market depends on an efficient transport system. The final breakthrough of the fossil-fuel-based regime would be industrial agriculture that directly uses non-solar energy for fertilization and as a source of power to intensify agricultural production. None of our cases fall within this last category.

In the following paragraphs, we seek to check the assumptions embedded in Table 6.1 by comparing our three cases according to various indicators relevant for our theory.

Patterns of Agricultural Intensification

Terrestrial ecosystems may be colonized in various ways. One class of intervention is landcover change: removal of forest in favour of agricultural and settlement areas and pastures. Another class of intervention is harvesting. Both classes of intervention typically result in a reduction or withdrawal of biomass energy from natural ecosystems in favour of its utilization by humans. As a summary indicator for the intensity of intervention we use HANPP, 'human appropriation of net primary production' (Haberl et al. 2007). Table 6.2[11] presents the major components of this indicator system for our three cases. NPP_0 gives the potential amount of energy produced through photosynthesis by green plants in a given territory in the absence of human intervention – this depends of course on the size of this territory but also on the natural vegetation that would prevail under given climatic conditions. HANPP amounts to the

Table 6.2   Colonization of land in the three cases: HANPP, harvest, and colonizing efficiency

	Nalang	SangSaeng	Trinket
NPP_0 terajoules/year [TJ/yr]	640	36	1090
HANPP [TJ/yr]	343	28	234
HANPP [% of NPP_0]	54	77	22
Harvest [TJ/yr]	24	11	7
Harvest [% of NPP_0]	4	30	1
Efficiency [$NPP_{harvest}$/HANPP] [%]	7	39	3

difference between NPP_0 and the biomass energy remaining in the terrestrial ecosystems after harvest;[12] it can also be presented as a percentage of NPP_0. The higher this percentage, the more intensive the sum of colonizing interventions. We may also calculate an indicator of colonizing efficiency: assuming that the key purpose of interventions is to secure a harvest to feed humans and animals, provide fuelwood, fibres and other products for human use, we may say the colonizing process is most efficient if it provides a maximum of harvest for a minimum of reduction of natural energy flows. This is certainly not a traditional agricultural efficiency standard, but it could be considered as a sustainability oriented standard: providing a maximum of service to humans for a minimum of damage to ecosystems.

What we see is an enormous span of intensity of intervention, which accords well with our assumptions as laid down in Table 6.1: the foraging community of Trinket appropriates only 22 per cent of potential biomass energy,[13] harvesting no more than 1 per cent. This low harvest is also the cause of the low colonizing efficiency, since the prevention of NPP through (hardly used) grassland is significant. In case this grassland were to be considered natural, as is sometimes argued, HANPP would fall even lower and efficiency would rise. Nalang comes next in HANPP with 54 per cent, followed by SangSaeng with 77 per cent – certainly a very intensive utilization of terrestrial ecosystems. This ranking complies perfectly with the presuppositions guiding Table 6.1 and the idea of intensification of use increasing from stage to stage.

Before we look more closely into the efficiencies of ecosystem use, let us scrutinize the conventional parameters for agricultural efficiency: yields per unit area and yields per inhabitant of the communities (Table 6.3).[14]

Table 6.3 Annual yield and labour time per agricultural area and per member of the community, and fossil energy use

	Nalang (rice)[a]	SangSaeng (rice)[a]	Trinket (coconuts)[b]
Area productivity [GJ/ha/yr]	30.7	21.6	104.1
Production per inhabitant [GJ/cap/yr]	7.1	18.4	7.6
Labour time per inhabitant [h/cap/yr]	304	486	14
Labour time per area [h/ha/yr]	1534	570	193
Fossil fuels in local energy consumption [%]	1.5	8.3	22.0

Notes:
[a] Production of paddy rice (preparation, transplanting, harvesting) only.
[b] Production of copra (maintenance, harvest, drying), then exchanged for rice, only.

With SangSaeng, we see a still traditional agrarian system driven to its limits, both in terms of its land and of its people. SangSaeng uses almost 80 per cent of its land as permanent paddy fields in order to produce rice both for subsistence and to generate cash income. The remaining 18 per cent constitutes (heavily degraded) forest. There is hardly any grassland, since shifting cultivation was abandoned long ago on the flat land. Livestock is left to graze on harvested fields and in the low-density forest (see Boserup 1965, where this phenomenon is described). HANPP amounts to 77 per cent of the total potential energy in biomass, and a high proportion of it constitutes harvest. On the other hand, yields per hectare are the lowest of the three cases, which has some climatic reasons (aridity), but possibly also social reasons: this community has stretched its labour power to the limits. On average, each inhabitant of SangSaeng (including babies and old people) works almost 500 hours annually for rice production (see Table 6.3) – this is substantially more than in both other cases. These 500 hours are spread very unevenly over the year: most of them occur in the wet season and as it approaches, people already dread the enormous workload that lies ahead of them (Grünbühel et al. 2003).[15] So SangSaeng cannot move further on the Boserupian path of investing more labour time per area to improve its yields. It has more or less reached the end point of traditional agrarian development. But why should this system be regarded as 'traditional'? In its direct-use patterns, it still relies mostly on solar energy – most of its fossil fuel use is related to transportation. On top of this, there is an 'indirect' energy subsidy: this subsidy consists in the provision of Thailand's infrastructure and transportation system, without which SangSaeng could neither produce rice for the market on the scale it does, nor relieve the pressure upon its territory created by part of its population migrating to the cities for work.

Nalang, in contrast, still retains some potential for better use of its natural resources. As a traditional mixed-type agriculture (Adriaanse et al. 1997) with predominantly subsistence production and hardly any use of chemical fertilizers and pesticides, most of the agricultural activity takes place on permanent paddy fields, although other areas are also used for shifting cultivation and grazing. Grasslands and secondary forests show high rates of NPP production since natural succession is more rapid than in primary forest areas. Nalang still maintains a dense forest area covering about 43 per cent of its territory. The secondary forest area accounts for 31 per cent of the total land, while grassland amounts to 10 per cent. Only 8 per cent of the total land available is used for permanent agriculture. HANPP is high (54 per cent) but since most of the NPP is prevented rather than harvested as a consequence of shifting cultivation, the colonizing efficiency is rather low (7 per cent). Furthermore, permanent agriculture and livestock herding are not as input-intensive (in the form of fossil fuels, fertilizers and animal feed) as in SangSaeng. The system does not run into a sustainability problem because the land area available is large and the ecosystem is very productive.[16] On the other hand, the system appears to make a fairly inefficient use both of its land and of its people: for an HANPP of two-thirds of SangSaeng, they produce only one-quarter of its harvest; for almost three times the labour input per ha they fail to achieve even 50 per cent more yield than SangSaeng (despite a more favourable climate), and while in total they work far less than SangSaeng's inhabitants, they achieve much less yield (see Figure 6.1) per hour of labour input. This is connected, as we discuss below in more detail, with the fact that there is as yet neither so much pressure nor so much temptation to be efficient:[17] Nalang's population density is low enough for the inhabitants to have little difficulty in feeding themselves and they have little opportunity to sell a possibly higher surplus on markets or to make use of labour opportunities beyond agriculture because of their remote location and the lack of adequate transportation facilities.

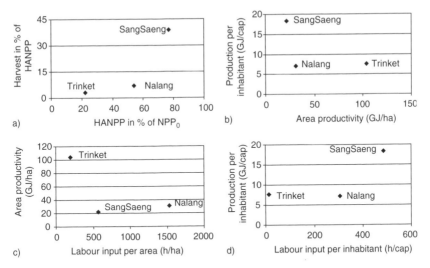

Figure 6.1   Various measures of efficiency of land and labour use for the three communities: a) land use efficiency (see Table 6.2); b) production per inhabitant in relation to area productivity (see Table 6.3); c) area productivity in relation to labour input (see Table 6.3); d) labour efficiency: production per inhabitant in relation to labour input (see Table 6.3)

So while SangSaeng and Nalang seem to share a development path on which Nalang can be seen as more 'backward' and SangSaeng as more 'advanced', Trinket seems to be a special case. The island in the Nicobar archipelago is sparsely populated but in former times, 32 per cent of the island's forests were converted into grassland.[18] The grasslands are maintained as grazing grounds by cattle and the inhabitants set fire to them in the dry season. Trinket's HANPP is by far the lowest of the three cases. Most of the land available on the island is either untouched or only used very extensively. While the coconut plantations cover only 0.7 per cent of the total land (and, traded for rice, provide the staple food needed), 55 per cent is still covered by primary tropical forest. In addition, a small amount of land is covered by mangroves, beaches and settlement areas. In terms of land and labour utilization, the situation on Trinket Island appears almost paradisiacal: the yield of the coconut plantations per unit area is very high, and the labour input needed is very low. So the inhabitants need to put much less strain on their land, and have the same returns per capita as Nalang, but for much less work. One might interpret this to mean that they make optimal use of the fossil fuel subsidy, namely in shipping copra for trade, which they exchange favourably for a staple food that would cost them much more land and much more labour if they had to produce it locally. This allows them to continue to lead a leisurely foragers' life (Sahlins 1972), to care little about efficiencies and to devote time and resources to elaborate ceremonies and pleasurable activities (such as fishing).

Self-sufficiency and Market Integration

Another standard for comparison across the three case studies considers the degree to which these communities are self-sufficient, and what role markets play for them. The percentage of food consumed that was produced locally may be used as an indicator of self-sufficiency (Table 6.4).

All of the cases studied show levels of consumption above the recommended minimum of food intake per day (2500 kcal/day; www.fao.org). These even lie above (albeit only slightly) the 2697 kcal/day average food supply for 'Developing Asia' (data for 2002; www.fao.org). In this sense, we may classify all three communities as quantitatively well-nourished. The production system can provide a sufficient amount of food for the population – at least in non-crisis years.[19] The adequate provision of food is achieved using different strategies. While Nalang manages to provide all the nutrition required from own production, the percentage of dependency on externally produced food is highest in the Nicobar/Trinket case (see Table 6.4). SangSaeng procures one-seventh of its food from outside. To interpret

Table 6.4   Consumption of food and percentage of imported food

	Nalang	SangSaeng	Trinket
Food consumed	2700 kcal/cap/day	3100 kcal/cap/day	2876 kcal/cap/day
Percentage imported	0	14	29
Rice consumption	277 kg/cap/yr	233 kg/cap/yr	64 kg/cap/yr
Percentage locally produced	100	100	0

this data we have to look at the specific economic structure of the three communities.

Nalang is endowed with sufficient resources of land, forest, water and wildlife. Its population relies on local resources because 1) they are available, and 2) trade is not a secure option due to the inefficient transport infrastructure in place. The staple food is rice, which is consumed three times daily, while additional food is hunted, gathered and raised in small kitchen gardens. Land resources in Nalang currently allow for sufficient rice production, including a small surplus, which is stored until the next harvest in case the next year brings underproduction. In this way, the staple is guaranteed in most years, whereas the availability of additional food items may fluctuate. Due to the high diversity of edible plants and animals available and due to the lack of major food taboos in the belief system of the Lao, additional food is always accessible, even if production varies. Hence, food production in Nalang remains at a low level, mostly relying on vegetal biomass and fish, but the yearly supply from domestic sources is even and secure.

Due to its geographic location and being a part of economically more highly developed Thailand, SangSaeng's food production system is more market-integrated than that of Nalang. Although rice, as the community's staple crop, is still produced locally (see Table 6.4), additional food items are purchased from the market. Instead of rice, processed noodles are consumed in rare cases, industrial condiments and spices are invariably applied, such as MSG and fish sauce, and meat is sometimes obtained from the market. Resources in SangSaeng have been degraded with the expansion of cropland. Money is available to the community via the production of cash crop rice, the activities of migrant labourers and the marketing of commodities in rural areas. At the same time, cultural change has taken place through the experiences of migrant workers in the cities and their interactions with people from other regions in Thailand. Marketing reaches the rural areas via the mass media and by way of trade networks. The transport of commodities is relatively inexpensive and time-efficient due to the extensive network of paved roads in Northeast Thailand. SangSaeng thus enjoys a relatively high level of nutrition (mainly due to the ingestion of meat) but increasingly depends for this on external supply chains.

Trinket's food system works differently. For historic reasons, the local economy of the island has relied on trade in copra. Traditionally, food was produced through exploitation of local resources, that is, fishing, gardening, forest product collection and livestock keeping. The staple used to be the starchy pandanus, which grew in the wild but was difficult to prepare. Gradually, over the past century, imported rice has replaced the indigenous source of carbohydrates. Today, copra is traded mainly for rice (and fossil fuels) mostly under a scheme of subsidized prices that buffer the fluctuations of the market. This scheme is part of the welfare programme implemented by the Indian government for the islands. Rice consumption is relatively low because it does not have the same importance as a staple food in the case of Trinket Island as in the previous cases. Rice is consumed frequently but not invariably and does not constitute the basic component of every meal. Instead, the diet relies heavily on fish, roots and tubers collected in the wild or from gardens, and coconut, a fruit of high energy content.

The percentage of food that is exported for sale at markets is yet another indicator giving evidence of the degree to which these local economies are self-contained and secluded. Alternatively, the extent to which they are integrated into markets – whether they sell only the surplus of what is locally consumed, or whether they intentionally produce crops or varieties specifically for sale (see typology in Table 6.1) – may be considered.

Table 6.5 depicts Nalang as a classical subsistence economy. The largest part of the agricultural harvest is consumed within the community, with only a minor part leaving the village for commercial purposes. In fact, rice is only sold beyond the village after internal needs have been met. The traditional system of storage keeps the rice harvest in the granaries until shortly before the following harvest. If, after one year, the rice supply has not dwindled due to household consumption, granaries are emptied in preparation for the next harvest. The amount left over is then sold off in order to make a little money to buy tools and fuel. The emphasis hereby

Table 6.5   Harvest exported to markets. Percentage refers to total harvest (crops, rice, livestock)

	Nalang	SangSaeng	Trinket
Exported crops [%]	18	52	49
Exported rice [%]	11	55	0
Exported livestock [%]	5	30	0

lies on meeting nutritional needs within the community and not on market production. On the other hand, the households of Nalang experience the pressing need to gain money in order to pay for taxes, agricultural inputs and petty commodities. They have reacted to this pressure by adding other crops to their production system. Cucumbers are raised in rice fields turned into gardens during the dry season. In what used to be the slack season, during which the emphasis lay on hunting game and gathering bamboo shoots, cucumbers are now planted in small-scale cultivation and sold at nearby markets. Similarly, banana cultivation has expanded from the traditional few plants to larger plantations. Both cucumbers and bananas are easily transported and survive several days of storage, should transport be interrupted due to lack of fuel or other problems. Thus, the two crops serve as an add-on to the traditional farming system and present themselves as an adaptation to encroaching market demand in the area.

SangSaeng has gone much further and has modified its rice production system. Whereas Nalang has not extended the agricultural area to gain money, its counterparts in Northeast Thailand have stretched the cultivation of available land to its limits. Nearly the entire area suitable for paddy rice production has been brought under cultivation. While the Lao community has invested in additional dry-season cultivation, this strategy is not feasible in Thailand's dry Isan plateau. Instead, the inhabitants invest more labour during the favourable wet season and plant additional rice to be sold on the market as a cash crop. The traditional staple in ethnic Lao communities, glutinous rice, is consumed by hardly any other culture on a regular basis and neither mainstream Thai nor the world market appreciate this variety. In order to adapt to market demand, SangSaeng inhabitants thus produce white 'Jasmine' rice, the sale of which secures them market prices. Half of the community's agricultural production is destined solely for the market and is not consumed by local households (see Table 6.5). The other half remains the subsistence production of the traditional glutinous rice. Similar to the Lao case, the rice produced for home consumption is geared towards meeting nutritional needs, with a slight surplus in case of exceptionally bad years. Here also, the surplus is stored until the next harvest and sold only if the amounts produced can easily fulfil local nutritional demands. 'Cash crop' rice is of a different variety and is subject to the law of external markets. The varieties planted require fertilizers and are supported by the agricultural administration. In addition, livestock rearing has gained importance in recent years. Buffalo and cattle not only serve their traditional purpose as working animals but are now kept to store and increase wealth. In times when immediate cash is required, livestock is sold on regional livestock markets. This practice has become an ancillary source of income for some of the more wealthy households in SangSaeng.

On Trinket, half of the coconuts produced are exported in the form of copra. Nevertheless, it cannot be claimed that planting coconut trees is a specialized production for the market since the very same coconuts are also consumed domestically. Nonetheless, it is also untrue to say that they only sell their surplus production – Trinket's inhabitants have to harvest substantially more coconuts than they directly need, as they rely entirely on this single product in order to acquire basic goods in return (rice, cloth and fuels). All other production in Trinket is purely subsistent, that is, it never leaves the boundaries of the island. Garden produce, fish and food gathered from the forest are mostly consumed by the household that collects them. Pig rearing – conducted mainly for ceremonial purposes and not for basic nutrition – experiences a limited amount of exchange. The flows, however, are minimal and do not serve an economic purpose. Rather, the exchange of pigs is a reciprocal arrangement involving village solidarity and ritual. One may assume that the number of pigs leaving the island is balanced by those received. The subsistence society has experienced adaptation to changing economic environments. While production of food and traditional exchange have remained intact to a large degree, modern-day needs are satisfied by way of selling copra, a product demanded by the industry. This arrangement allows Trinket inhabitants to become mobile (by obtaining fuel for boats) and to enjoy high-quality food (rice), which they would otherwise lack, or for the production of which they would have to invest much labour.

6.4   DRIVERS OF CHANGE

Population Dynamics and Density

For Boserup (1965, 1981), population dynamics and, as a consequence, population density, are the major drivers of agrarian innovations and social change. According to our data (Table 6.6),[20] SangSaeng has by far the highest population density, as should be expected, whereas Nalang shows a figure of less than a half. Trinket's population density is the lowest, but

Table 6.6   Population density and population growth

	Nalang 2001	SangSaeng 1998	Trinket 2000
Population density [cap/km²]	43	93	11
Population growth [%/yr]	3	1.4	1.5

with 11 persons per km^2 is still much higher than to be expected from a 'pure' foraging society.[21]

SangSaeng's population growth is moderate (possibly already as a result of out-migration), and as we have seen, it could use more labour power. But as the local resource base is more or less exploited to the limits, a larger population that might lessen the burden of work might also reduce food security and income. It appears likely that SangSaeng will not be able to sustain its growth rate. Already, the social system has to engage in a temporal migration strategy in order to supply labour force when needed and reduce the workforce during the slack season. In addition, permanent migration, which is a traditional cultural feature of Isan society,[22] would need to increase gradually in order to keep the population balance of the village stable (see Fukui 1993).

In Nalang, limited cultivable land coincides with high population growth. The (Boserupian) response is to invest labour into land and achieve higher absolute yields. It is most probable that this process will lead to a qualitative transition of the farming system in the near future, for example, by increasing the existing dry-season cultivation through investment in irrigation schemes. There should be enough labour reserve for this strategy. Nonetheless, with a growth rate of 3 per cent annually, the local population will within decades outgrow the community's resource base[23] and out-migration will be required to stabilize or reduce population. Trinket as a foraging society has vast land resources available. To an even greater extent, the ocean provides a rich source of proteins and energy for nutrition. There seems to be no problem in maintaining a relatively high density (by foraging standards) as long as the inhabitants have access to rich marine resources and – owing to their geographical location – are able to procure additional food and commodities by trading in coconuts with passing vessels. Thus, the Nicobarese are not entirely dependent on seasonal availability of food since there are abundant coconuts that can be traded. Moreover, given that there are still uninhabited islands in the Nicobar archipelago, population pressure is kept under control by a strategy of expansion to other, still uninhabited islands of the archipelago. Trinket itself was inhabited no earlier than 150 years ago and as a result, has a population density of less than the average of 23 cap/km^2 in the Nicobars in general. This means that Trinket could still double its population density without facing major resource constraints, and, were such constraints to develop, the population could expand to another, as yet uninhabited, island.

The decadal population growth on Trinket during the last 100 years has been rather erratic, dropping at times to minus 21 per cent between 1911 and 1921, and rising to plus 151 per cent in 1981. Such fluctuations can either be explained by outbreaks of epidemics[24] or by immigration, the latter being rather common, and by the fact that the population of the Nicobars in general has been increasing at an average rate of 20 per cent per decade.[25] According to the Boserupian hypothesis, the strategy of expansion for as long as possible precedes intensification. Given that there is still plenty of space for expansion among the Nicobar Islands, this strategy has not yet been exhausted. However, exposure to a different economic paradigm introduced by foreign traders and the government, as well as the increasing needs of the Nicobarese, might expedite the process of expansion and eventual intensification.[26]

The inhabitants of Trinket, as described above, need to invest very little working time in order to produce for external markets. They also spend very little time meeting subsistence needs through domestic production. The Nicobarese spend an average of 14 hours/cap/yr on the production of marketed copra and 121 hours/cap/yr fishing their coastline for food. This leaves them with plenty of leisure time. Müller-Herold and Sieferle (1998) discuss 'leisure preference' as a risk avoidance strategy for societies under fluctuating environmental conditions. While a complete utilization of all resources (optimizing production) results in population growth and makes the society dependent on high consumption levels, the under-use of resources maintains a safety margin to avoid 'ruin' (Müller-Herold and Sieferle 1998). However, at present, with low population density, relatively low population growth and rich land and water resources for their exclusive use, the Nicobarese seemingly have little to be worried about. They do, however, see the gradual encroachment of settlers and traders, with different needs and consumption patterns, as a danger to their lifestyle – as well as government welfare schemes that put them under pressure to adopt higher levels of market production and consumption. Their concerns may be well-founded, since although their production system is well adapted to its environmental conditions, it is vulnerable to external interventions that may induce major qualitative transitions regardless of the will of the Nicobarese.

Access to Markets

In the case of the Nicobarese, we can see that access to markets may exist for a long period of time – in this case centuries – without affecting the basic structure of the subsistence economy (quite in contrast to the key thesis of Polanyi 1957). The location of the islands along a historically important trade route attracted passing vessels that anchored there to replenish food and water supplies. However, for the Nicobarese, this form of barter-trade came to be embedded in their socioeconomic fabric. It could be said that the market came to the Nicobarese rather than the other way round. The Nicobarese continued to live like hunter-gatherers-cum-horticulturalists with very little time – 20 minutes/cap/day – invested in economic activities (Singh et al. 2001). Nonetheless, trade with the outside world is essential to their survival. Over the last 250 years, the Nicobarese economy has been well integrated into the regional and even global economy, insofar as a decline in copra prices elsewhere has had a great impact on the Nicobarese. Since the mid-1980s, the copra industry has twice faced crisis, once in 1987 and again in 2000. In both cases, the Indian government permitted the import of coconut oil from other Southeast Asian countries in response to high copra prices within the country. Consequently, the Nicobarese faced impoverishment and sought government intervention in the form of a Copra Price Support Scheme. Once established, the scheme was influenced by private traders on the islands, who found a way to make even higher profits at the cost of the growers. Facing great problems, the Nicobarese in 2002 decided to go to court against the Indian administration for mismanaging indigenous lands by allowing private traders to settle in their territory (which is banned under the Protection of Aboriginal Tribes Regulation of 1956) and for the failure of the copra support scheme. The case is still pending and the Nicobarese meanwhile strive to maintain the connection to the global market, on which they depend so much, and also to seek economic options other than copra to meet their increased needs.[27]

In other cases, new market opportunities may dramatically change the economic outlook of the community, as is the case in the historical development of Northeast Thailand where major national efforts have been made to integrate the region into mainstream Thai society. Infrastructure development (rail, roads, telecommunication and electricity) has brought markets into the vicinity of local communities. Not only has access to markets been facilitated through these investments, but new markets have been created via social networks and access to information, creating new demands. In this way, the Thai government has successfully promoted cultural mainstreaming and reduced the cultural and the economic distance of remote areas from marketing. From a theoretical point of view, looking at biophysical change, such infrastructural efforts appear in the form of an indirect energetic (fossil fuel) subsidy to otherwise solar-energy-based rural communities. Without such a subsidy, they could not and would not have been able to access markets.

For SangSaeng, this 'subsidy' allows a way out of what would have become a deadlock involving resource over-utilization and finally pauperization. As land has become increasingly scarce due to the constant expansion of land under agriculture, even small-scale population growth places the subsistence production under pressure. In addition, subsistence is endangered by the scarcity of other resources. While the production of rice is still sufficient, other foods cannot be produced in adequate quantities. Forests have been degraded heavily and the climate prohibits additional agricultural production during the dry season. This has forced the villagers of SangSaeng to enter the market. As a first step, they have switched to cash crop rice in order to generate cash income needed for inputs, such as fertilizers, taxes and fuel. On the other hand, this intensified production for outside markets also increases the pressure upon land and labour power.

In the dry season, SangSaeng villagers now also participate in the labour market to raise cash for their families at home. Almost 20 per cent of the population engage in seasonal labour migration. In addition, young adults move to the cities before marriage with the aim of saving funds for future establishment of a household in the village and to pay bride-wealth (dowry) when marrying. Temporal migration of this kind is an additional factor. Not all children can count on the existence of an inheritance. Usually, the youngest daughter, who takes the responsibility for supporting her parents in old age, is granted an inheritance. Other siblings have to move elsewhere and sometimes move out of agriculture into the city. In addition to seasonal and temporal migration, we can observe permanent migration to other villages and to commercial centres. The stretching of resources has reached its limit and the system can only be maintained if there is a transport system enabling seasonal, temporal and permanent migration to participate in urban labour markets (see also Fukui 1993).

In the case of Nalang, we find that the market plays a minimal role. This owes much to the fact that the village is 180 km away from the closest economic hub and is located in a valley with no major transportation thoroughfare. External influences became more prominent with the construction of a road in 1980 and consequently integration with the market became more pronounced. Interestingly, market integration, although still minimal, has mostly taken place through the introduction of a few consumer goods and food items to the village. Major items of processed food available in Nalang today include ice for cooling, soft drinks, beer and fish sauce. Consumer goods and agricultural implements purchased via markets include motorized ploughs, cloth, radios, bricks and sheet metal. On the other hand, only a minimal portion of the agricultural produce is exported, while most is used for subsistence, including livestock. The acceptance of technology in the form of the motor-plough, a consequence of government policies, did little to contribute to the village's integration with the market. On the contrary, it provided a solution to the need to intensify agriculture for subsistence arising from the increased population density due to immigration.

While the existence of markets certainly plays an important role in driving transitions of the three local economies discussed here, the process and timing of integration with markets takes a different form in all three cases. In SangSaeng, Thailand, market integration was achieved by the government via a deliberate development strategy that promoted transport and communication infrastructure, chemical fertilizers and hybrid seeds, small loans for farmers, and income-generating cash crop production. In Nalang, a similar process has taken place but at a much slower rate and not via national development strategies. Rather, the development of road transport was enforced by semi-private logging companies who themselves required roads for timber extraction. Similarly, a system of loan-giving developed through private initiative. Farmers in need of cash sell their crop before it is harvested and pay back the loan in kind when the season arrives. The system is not cushioned by state regulation and thus many farmers are drawn into debt through having sold more crop than they have actually produced. As a consequence, consumption is not promoted but instead remains at a low level, since capital is not readily available. Due to the favourable location of Trinket Island, the Nicobarese have had a long history of market contact with passing ships. This has allowed them to control their own resources. Although consumer goods are increasingly purchased, the state-controlled system of exporting copra at subsidized prices has protected the Nicobarese from having to adapt too quickly. On the other hand, the gradual shift in consumption from home-grown pandanus to imported rice (traded for copra) as a staple diet represents a decrease in food security for the Nicobarese. They are dependent on the state-controlled system and would thus be at the whim of the government should it ever decide not to continue the trading scheme.

Trinket represents a somewhat different case from the other two under discussion in this chapter since it does not rely on agriculture for subsistence. The type of horticulture and coconut plantations managed by the inhabitants does not consume large land resources. Moreover, sandy beaches are preferred locations for coconut plantations as they mature in about seven years (compared with 15 years when grown inland) and do not require much care, for example, in the form of watering. To date, the coconut (and from this, copra) production has proved sufficient to provide the rice staple in exchange, while fishing produces a large portion of the Nicobarese daily diet. Thus in the case of Trinket, the usage of land resources is not comparable to that required by agricultural societies.

State Intervention and Development Programmes

We have described the form of state intervention that has led to the far-reaching integration of SangSaeng's subsistence community into the nation-state, as a result of which the community is embarking on a different path of development. In a deliberate effort to develop Northeast Thailand, infrastructure and communication networks were built, educational programmes were introduced that enforced cultural mainstreaming (nationalism, standardized language, Buddhist values), and the system of public administration was homogenized. The programme to develop Northeast Thailand was introduced for political reasons as well as for reasons of poverty alleviation and food security. Northeast Thailand had once been the primary hotbed of government opposition and the base of the Thai Communist Party. Economic development was seen as the way to eradicate political adversaries (Keyes 1967). Until today, forms of grassroots political opposition have been quite successful as exemplified by the 'Assembly of the Poor' movement, which was powerful enough to enforce the abandonment of the Pak-Mun hydroelectric dam by the government. Similar development programmes are currently underway in the south of Thailand aimed at quelling the Muslim resistance movement.

Despite political resistance, the development programme undertaken by the Thai government has served to integrate the Northeast Thailand area as a 'Thai' region.[28] Villagers in SangSaeng respect the king as the patron of the country, they perceive their (Lao) language as a regional dialect not to be spoken in official settings, they accept their low position in Thai society and the gradual degradation of resources that they have relied upon for centuries. Due to the absence of natural barriers, Northeast Thailand was never an isolated area but was always seen as a hinterland, both for the Thai and for the Lao kingdoms. Population densities remained low until the 1960s and cities remained administrative outposts of the government without particular relevance for most sectors of the population. The transition has thus taken place over a mere four decades and is still ongoing.

Nalang has not yet experienced any major national development programmes. Transition was initiated by the construction of a logging road through the valley, which in its wake brought merchants, money lenders and the possibility of selling crop. State intervention is confined to the village headman serving as a link between the community and the national administration, although he is chosen by the community. Land-use regulations were introduced recently and there are indications that these are actually being enforced. These constitute an additional land-use constraint upon the farmers but do not lead to a major transformation of the agricultural system, which remains based on shifting cultivation and rain-fed permanent farming for subsistence production. The government appears to refrain from intervention as long as the community is able to produce sufficient food. The gradual increase of consumer items and processed food is not state-enforced but rather a consequence of the logging road's existence. Nonetheless, needs and wants are on the rise and first indications from a survey of a few households show that the villagers wish to continue consuming goods to which they have newly acquired access. In order to do so, they require more money, which will lead them to intensify cash cropping and (illegal) extraction of forest products. This represents a vicious circle in which their resources will be drained and migration will be the ultimate consequence.

In 1947, the Nicobars became part of independent India. In order to offer protection to the islanders, and in recognition of the islands' strategic military location, the Protection of Aboriginal Tribes Regulation (1956) strictly regulated entry to the Nicobars. More recently, since the 1980s, efforts have been made by the local administration to integrate the islanders into mainstream society by introducing tribal welfare measures, such as education, health care, communication facilities and so-called 'economic upliftment' programmes. This entailed the construction of new buildings throughout the archipelago, not only for infrastructure purposes, but also to provide housing for administrative officers. New labour was brought in for construction work and the island's population grew substantially. By the 1990s, these labourers and spouses of government officials took advantage of the opportunity presented by having been allowed into the islands and set up small enterprises. Small markets were established where the Nicobarese could sell their copra and obtain food and consumer goods. The availability of manufactured goods and increased communication led the islanders to aspire to a lifestyle comparable to that of mainland India. These needs had to be met by increasing copra production as well as incurring debts with the traders.

6.5   CONCLUSION: 'DEVELOPMENT' OR 'SOCIOECOLOGICAL TRANSITIONS'?

This chapter has presented three case studies on rural communities from Southeast Asia that are undergoing drastic changes. While analysing each of those cases according to its very particular features and circumstances, we have tried to put them in a certain order to enable us to pursue our theoretical question: how can the process of change from hunting and gathering across agrarian subsistence regimes to integration into an industrial, fossil-fuel-powered society be described? Is this a smooth, incremental and gradual modernization process, or are there thresholds at which communities face major qualitative change? Is, indeed, the energy regime the key condition to be aware of? How do system levels interact, that is, how do urban markets and national state interventions impact upon processes at local level? Finally: what role do natural resources and conditions, and their modification through human activity, play in moulding the course of this process? These are very wide-ranging questions but, nevertheless, we shall attempt to make a resumé of our findings in this chapter in the light of these questions.

According to classical development theory, development is connected to economic growth, which is, in turn, associated with a gradual increase in the quality of life. Development, in this sense, reflects gradual changes in the economy of the society, inclusion of modern lifestyles and consumption behaviour, modernization of social institutions and mech- anization of the economy. From what we were able to determine, the difference between Nalang and SangSaeng can be generally interpreted in this way. Nalang is more 'backward', and SangSaeng is more 'advanced', using the same set of parameters. This is compatible with our theoretical assumptions: these two communities are both subject to an agrarian socioecological regime, with one of them having moved further through the life-cycle of this regime. On the other hand, if we look at a wider range of variables than are considered in classical modernization theories, we can see the more advanced of these two communities is in fact close to a deadlock, to an 'end-of-the-road' situation from which point it will not be able to continue as it did before. This end-of-the-road situation has much to do with the exhaustion of natural resources (including human working time) and can only be overcome by intervention from a higher system level (state and/or market), drawing its momentum from a different source of energy.

In both these cases, the limiting factors are indeed productive land and energy, the source of which is predominantly biomass. In SangSaeng it is neither possible to bring more land under production, nor to intensify further in the absence of fossil-fuel-based agriculture. SangSaeng has managed to get out of its resource constraint by generating additional income from off-farm labour. However, as population grows and needs increase, new possibilities would have to be explored. One likelihood would be to turn into a full-blown fossil-fuel-based market agricultural economy and bring the entire available land under production for the market. In sub- stituting human labour with fossil fuels, future availability of human time would increase, which would, in all likelihood, lead to enhanced migration into the cities in search of wages. Indeed, relatively easy access to the market has been a crucial factor for driving SangSaeng to its limits, and in all probability may be equally responsible for a transition to fossil-fuel- based agriculture.

Nalang, on the other hand, is still able to stretch its capacities, both in terms of land and human labour, but the temptation to do so is lacking since its people's subsistence needs are already being met with relative ease. Population density is half of that of SangSaeng and so the pressure to increase productivity is not so high. They can even afford to be extravagant with their labour time, investing three times more per hectare than their counterparts in SangSaeng with only a one-third increase in productivity. Had Nalang had easy access to markets like SangSaeng, the probability that they would have already started to produce rice for the market earlier would have been higher. At present, production for the market is limited only to cucumber in the dry season, which is a minor crop. Thus, Nalang, though at present still an agrarian regime like SangSaeng, does not face a dead-end situation and has the chance to increase its population and optimize available natural and human resources without the need to alter production methods dramatically.

Among the three cases, Trinket stands out, not fitting the axis of 'progress' or 'development'. Combining features of traditional foraging societies (namely, an almost extravagantly leisurely lifestyle in combination with a rich and varied diet) with a well-organized form of market utilization, Trinket still has the possibility to stretch its capacities both in terms of land availability as well as human resources. A low population density coupled with rich forest and sea resources are excellent conditions to ensure sufficient nutrition for the inhabitants the year round. The efficient exchange rate between copra and their staple – rice – further contributes to their security net. Were the Nicobarese to produce their own rice, they would require at least three times more land than they currently need to produce the equivalent of coconuts. Indeed, their favourable conditions can be attributed to their location on an important sea lane, and thus to the market coming to them (instead of the converse, which is usually the case), as well as to the subsidies and security net provided by the Indian government. Consequently, the Nicobarese require little land and less working time, both of which are stretchable to a large degree, to meet their requirements. Thus Trinket does not appear to be on the same development path, with similar opportunities and constraints, as the other two cases.

While an analysis of three cases does provide relevant insights into local transitions in developing countries, it would be worth taking up a few more examples to be able to comprehend more accurately the socioecological transitions taking place at the local level. At the same time, studies that engage more scales of analysis would be more meaningful. As far as these cases go, we do not see a smooth, incremental and gradual modernization process but rather several factors interacting at the same time at multiple levels that sometimes push them into threshold situations instigating societies to undergo change.

APPENDIX 6.1   METHODS

As a first step, we specified the biophysical compartments of the social system under investigation, comprising the human population, domesticated livestock and durable artefacts (huts, buildings, wells, pathways, boats, machines and so on), together termed as society's stocks (or biophysical structures). The next step was to define the territory that the social system is entitled to exploit, that is, its 'domestic environment' (forests, agricultural fields, shores, sea zones and so on). It should be noted that while the 'territory' is a spatially explicit area of land that can be described geographically, its boundaries are not defined naturally but socially, as the area the local community under consideration is entitled (legitimized by political, legal or informal social consensus) to exploit (Fischer-Kowalski and Erb 2003). This may include areas (for example, sometimes forests) it proportionally shares with other communities. All materials (and energy) required to reproduce the biophysical compartments (stocks) of the socioeconomic system and therefore organized and channelled through social processes, in particular, human labour, are considered socioeconomic material and energy flows.[29]

Having defined this, we were able to make a distinction between two boundaries: one that distinguishes the society's stocks from the stocks in its domestic environment and another that separates the socioeconomic system from other socioeconomic systems. The two boundaries allow us to distinguish between two types of socioeconomic material and energy flows entering the system. First, the extracted and appropriated energy/materials from the domestic environment, which we term domestic extraction (the most important of these, of course, being biomass harvest), and second, imported energy/materials from other social units. Equally, on the output side we differentiate between 1) wastes and emissions that are excreted into the domestic environment and 2) material exports to other social units. Clearly, when balanced by ancillary flows of water and air, material inputs have to equal outputs.

For a stock account of the social system we made an inventory of the local population, livestock and the most important human-made structures. Subsequently, we attributed weight (metric tons) to these stocks. Whenever possible, this was done by weighing these artefacts. However, in some cases, we had to rely on factors from externally available sources. All other durable artefacts such as furniture, out boats, canoes, power generators, sewing machines and so on, were categorized by sizes such as large, medium and small, and average weights were attributed to each of them. All data concerning the society's material flows was collected in accordance with the key distinction between materials domestically extracted and traded materials as defined by the two system boundaries mentioned above. The unit for material flows used was metric tons/year. All biomass flows were calculated both in terms of weight of dry matter and fresh weight when harvested[30] or traded. To account for the energetic input and energy conversion processes, we used the same systems boundary as in material flow accounting (Haberl 2001a; 2001b). Material flow data for biomass and fossil fuels was converted to energy units by using calorific values. Having generated the above data, the following indicators were derived using the given formula:

1. *Weight of stock* (WoS): the total weight of human-made artefacts in metric tons;
2. *Domestic extraction* (DE): materials extracted domestically;
3. *Direct material input* (DMI): materials imported + domestically harvested;
4. *Domestic material consumption* (DMC): DMI – exported materials;
5. *Direct energy input* (DEI): energy imported + energy domestically harvested;
6. *Domestic energy consumption* (DEC): DEI – exported energy.

Land use was studied using the same logic as with defining the systems boundary. First, we accounted for the 'total area' owned and used by the social system to meet its metabolic requirements. As a next step, we calculated the area representative for the different ecosystem types (such as primary and secondary forests, grasslands, mangroves, horticultural gardens and agricultural fields) and their uses as well as an overview of the total landcover of the social system (including area under settlement and related infrastructure). Data for landcover and ecosystem types were either taken from official statistics (especially in the case of forest cover, grasslands, mangroves and beaches) or manually measured (as in the case of settlements, agricultural fields and coconut plantations).

Inextricably linked to metabolic exchanges between society and nature lie eco-regulatory strategies, whereby humans intervene in natural processes with the intention of transforming certain parameters of the ecosystem to enhance their utility. An example of this is agriculture. This has been termed colonization of natural processes (Fischer-Kowalski and Haberl 1993). In order to compare the intensity of colonization across the three local cases, we calculated the human appropriation of net primary production (HANPP). Generating data on HANPP is achieved in five steps and requires the calculation of: 1) the net primary production of the potential vegetation (NPP_0), or the amount of NPP that would prevail in the absence of any human presence or intervention; 2) the NPP of actual or current vegetation (before harvest) in the face of human-induced landcover

Table A6.1   An overview of land, population and livestock resources in the three cases

	SangSaeng	Nalang	Trinket
Area (ha)	184	1627	3626
Area (ha/cap)	1.1	2.3	9.1
Agricultural land (ha)	148.2	293.7	1185.7
Grains (ha)	146	135.9	n/a[c]
Gardens/Plantation (ha)	2.2	n/a[b]	25.4
Pasture (ha)	n/a[a]	157.8	1160.3
Forest (ha)	34	1322.6	2284.4
Residential area (ha)	0.8	13.4	0.4
Population (cap)	171	702	399
Population density (cap/ha)	93	43	11
Cattle (cows, buffalo – no.)	108	109	89
Livestock (no./cap)	0.6	0.2	0.2
Yield [GJ/ha/yr]	30.7	21.6	104.1
Yield [GJ/cap/yr]	7.1	18.4	7.6
Labour [h/cap/yr][d]	304	486	14
Labour [h/ha/yr][e]	1534	570	193

Notes:
[a] Pasture on rice fields during dry season. During the cropping season, cattle are fed by cutting grass along the field boundaries.
[b] Gardens are included in residential area.
[c] The Nicobarese purchase their staple in return for copra.
[d] Labour time invested in rice or copra production per total population.
[e] Labour time invested in rice or copra production per agricultural area.

change; 3) the annual biomass harvest; 4) the remaining NPP available in the ecosystem (current vegetation minus harvest); and finally 5) the HANPP by subtracting the remaining NPP (after harvest) from the potential NPP. (See Table A6.1 for an overview of land population and resources for Trinket, SangSaeng and Nalang.)

ACKNOWLEDGEMENTS

Research on the Nicobar Islands was initiated with the help of a field research grant to Simron Singh from the Government of India (Department of Culture, Ministry of Human Resource Development). Subsequent work was carried out by two research fellowships awarded to Simron Singh, namely the International Fellowship of the Austrian Ministry of Science, Education and Culture, and the START Visiting

Fellowship supported by the research programme IHDP-IT. Research on SangSaeng (Thailand) was financed by a grant from the Jubiläumsfonds of the Austrian National Bank. The study of Nalang (Laos) was part of a larger research project, 'Southeast Asia in Transition', funded by the European Commission's Fifth Framework Programme (INCO-DEV), which was led by the Institute of Social Ecology and supported by partners from Southeast Asia and various European countries. All these case studies were part of a flagship project endorsed by IHDP-IT, 'Transitions from an Agrarian to an Industrial Mode of Subsistence – and Beyond', directed by Professor Marina Fischer-Kowalski.

The authors wish to thank Heinz Schandl, Verena Winiwarter, Fridolin Krausmann and Helga Weisz for useful discussions and comments.

Previous publications on these case studies, with more methodological detail, include Grünbühel et al. (2003); Singh (2003); Singh and Grünbühel (2003); Singh et al. (2001).

NOTES

1. Occasionally, when the market price for copra drops below a reasonable level, the government provides relief to the Nicobarese by buying their copra at special prices.
2. Following the tsunami disaster of 26 December 2004, the islands faced an uncertain future. Being in close proximity to the epicentre, the islands were heavily devastated, with a third of the population killed and almost all of the coconut plantations in coastal areas washed away by the waves. While the Indian administration provides relief and rehabilitation, the Nicobarese resolutely emphasize the importance of providing food security themselves. Food-producing gardens consisting of a variety of yams, fruits and vegetables are being established, something that was hitherto uncommon. Part of the garden produce, along with fish, is expected to be sold for cash if favourable market conditions prevail. For some, off-farm labour, which is being offered as part of the reconstruction work after the tsunami, is seen as yet another option for cash generation. The more enterprising Nicobarese are considering starting micro-businesses on the islands, if their skills can be matched with market opportunities, or even migrating to the city of Port Blair (in the neighbouring Andaman Islands) in search of jobs, both of which options are rather exceptional.
3. For a more detailed discussion on data generation methods see Grünbühel and Schandl (2005); Grünbühel et al. (2003); Singh and Grünbühel (2003); Singh et al. (2001); and Appendix 6.1 at the end of this chapter.
4. Only 8 per cent of households raise cattle.
5. 43 baht = approx. 1 US$ in 2001.
6. The tsunami of 26 December 2004 reduced the Nicobarese population by nearly a third.
7. Copra is produced from dehydrated coconuts that are baked over a fire for 15–17 hours, after which oil is extracted from the meat and then used in several industrial processes such as soaps, cosmetics, paints and margarine.
8. The first mention of copra production in the Nicobars is found in the *Annual Report* (1883–84), though presumably it must have started a decade earlier.
9. In a more detailed argument than the one employed here, other limitations beyond energy would have to be considered too, mainly limited availability of freshwater and limited soil nutrients.

10. For an analysis of transport as a strong limiting condition of agrarian systems see Fischer-Kowalski, Krausmann and Smetschka (2004).
11. Data in Table 6.2 refers to Trinket according to Singh and Grünbühel (2003) and Singh et al. (2001) and are restricted to above ground compartment. It should be noted here that the study of Trinket applied a somewhat different definition of the social system than used for SangSaeng and Nalang. In the case of Trinket, an entire island rather than a single settlement was investigated. In consequence, the island's natural boundaries were assumed to be the boundaries of the social system. In practice, Trinket is the home of three villages, all of which maintain very close contacts with the neighbouring island. While the cases of Nalang and SangSaeng took the administrative boundary to define the social system, the Trinket case study used the natural boundary of the island. This complies with the definition of 'territory' nevertheless, since the land area was not contested by any other social system and was entirely used by the island's inhabitants. However, HANPP on Trinket does not include an aquatic component despite the importance of fish catch.
12. Harvest, as has been stressed before, does include animal grazing.
13. There had remained some doubt as to whether the grasslands on Trinket exist due to natural processes or are a result of colonization by Danish invaders. What is certain is that the colonizers left cattle on the island that continue to maintain the grasslands as such and prevent succession. In Table 6.2 it is assumed that the grasslands are human-made, yet HANPP still does not rise above 22 per cent, by far the lowest among the three sites. Newest results (see Singh and Grünbühel 2003 and compare with Singh et al. 2001) point towards the natural occurrence of the grasslands.
14. The data represented refers to labour time spent performing the most relevant agricultural activities in terms of livelihood generation. Nalang residents spend additional time in upland rice cultivation and SangSaeng residents spend time in gardening activities during the dry season. Trinket islanders produce copra in order to exchange the product for their staple, rice. They spend much more time on other activities, such as fishing.
15. In SangSaeng, labour demands are met by the production system in the wet season when rice is cultivated. Population is dimensioned in a way that the labour force can meet peak demand. The dry season, however, poses increased risk to the sustainability of the community since: 1) significantly less labour is required; 2) food availability (other than rice) decreases drastically; and 3) the increased need to gain money cannot be matched by the agricultural production. Thus, a fraction of the population (20 per cent) moves elsewhere during the dry season to find paid work. These migrant workers subsist on the rice produced at home but consume additional food (and, of course, use the freshwater) at their workplace and earn money, which they bring back to their households and thereby generate monetary income. The high labour demands of rice farming can thus be buffered by seasonal migration, which additionally serves as a strategy to gain money required due to the community's relatively high level of integration in the national economic system.
16. Precipitation is at least 700 mm/yr higher in Nalang than in SangSaeng.
17. The figures shown in Table 6.3 include labour in both paddy fields and swiddens. Although paddy production is very labour-intensive, in the observed case, swiddening involves even more work. National legislation and land zoning by the government (Mayrhofer-Grünbühel 2004; Thongmanivong 2004) have made shifting cultivation in Nalang an immensely inefficient activity. The social limitation of land has led to frequent intervals of land-use and as such does not leave sufficient time for land regeneration and natural succession. This has led to the development of weeds on the planted areas and farmers spend most of their time weeding the fields to allow the rice plants to grow. The current three-year cycle is unusual for shifting cultivation. Farmers are aware of this and wish to change to paddy farming, which is not possible due to the exhaustion of land capacities. Thus, shifting cultivation must be engaged in by those who do not own sufficient paddy land in order to meet nutritional ends. In view of the growing population and given the lack of other opportunities, the strategy also serves to engage the large amount of available labour power for at least a part of the year. In the long run, however, the strategy is not sustainable and the community will have to adopt other strategies that are more efficient and provide better yields.

18. See note 13.
19. In case of climatic turbulences or environmental fluctuations, the provision of food may be endangered. Nevertheless, all three communities have had previous experiences with catastrophic disturbances of the food system and have adapted to these through coping mechanisms and feedback loops. However, the tsunami disaster of December 2004 is an example of a total breakdown of the existing food system in the Nicobar Islands, with which they could not cope locally and as a result became fully dependent on aid.
20. Trinket data according to Singh et al. (2001) and Singh and Grünbühel (2003); Nalang data according to official district (MuangFuang) statistics for 2000; SangSaeng data according to provincial (Ubon Ratchathani) statistics for 1998. Sources of data on population growth: Trinket data calculated from decadal growth rate between 1991–2001 (Census of India 2001). For Nalang and SangSaeng official data is used and represent yearly population growth.
21. The average density of foragers in the neighbouring Andaman Islands is about 0.8 pers/km^2, which is still rather high if one looks at population densities of foragers living in tropical forests (Kelly 1995).
22. Temporal migration, like permanent migration, is a Lao/Isan cultural characteristic. Individual families and entire villages have traditionally engaged in seasonal migration during the dry season, for example, for fishing expeditions, salt extraction, sale of handicrafts (see Klausner 1993 and for a literary description EEA 2003). Permanent migration and partition of villages due to population pressure have been described by Fukui (1993) and also mentioned by Hanks (1972).
23. At the present growth rate, Nalang will reach SangSaeng's population density within 27 years. It appears likely that Nalang will not be able to sustain such a growth rate for long, since land suitable for agriculture has already become scarce and the strategy of expansion has more or less reached its limit.
24. The *Annual Reports* of the British administration between 1924 and 1940 are characterized by reports on their efforts to contain the outbreak of diseases and epidemics, namely venereal diseases, influenza, smallpox, measles and dysentery. The mortality rate was reported to be high during this period (Singh 2003).
25. This is similar to the decadal population growth rate of India, which was 24 per cent in the period 1981–91 and 21 per cent in 1991–2001.
26. There is a high probability that the economic transition might take place in the aftermath of the 2004 tsunami, which shattered the traditional economy and destroyed coconut plantations.
27. The situation following the 2004 tsunami is quite different, since copra cannot be produced on the island any more. This new situation is, however, beyond the scope of this publication.
28. Despite the development programme, inhabitants of Northeast Thailand remain at the bottom of the social pyramid and are mostly found occupying lowly positions in the cities, for example, as Tuk-Tuk drivers and housemaids.
29. Theoretical considerations have been raised as to whether plants should be included as part of the socioeconomic system or not (Fischer-Kowalski 1997). For pragmatic reasons, however, they have been excluded so as to take advantage of available agricultural statistics as well as avoiding complications on where to draw a boundary between the socioeconomic and the natural system (Schandl et al. 2002).
30. Livestock grazing was estimated and considered as part of the harvest.

REFERENCES

Adriaanse, A., Stefan Bringezu, A. Hammond, Yuichi Moriguchi, Eric Rodenburg, Don Rogich and Helmut Schütz (1997), *Resource Flows: The Material Basis of Industrial Economies*, Washington DC: World Resources Institute.

Andrews, Harry V. and Vasumathi Sankaran (eds) (2002), *Sustainable Management of Protected Areas in the Andaman and Nicobar Islands*, New Delhi: ANET; IIPA; FFI.

Annual Report (1837–38), *Report on the Administration of the Andaman and Nicobar Islands, and the Penal Settlements of Port Blair and the Nicobars*, Calcutta: Office of the Superintendent of Government Printing.

Annual Report (1883–84), *Report on the Administration of the Andaman and Nicobar Islands, and the Penal Settlements of Port Blair and the Nicobars*, Calcutta: Office of the Superintendent of Government Printing.

Boserup, Ester (1965), *The Conditions of Agricultural Growth. The Economics of Agrarian Change Under Population Pressure*, Chicago: Aldine/Earthscan.

Boserup, Ester (1981), *Population and Technology*, Oxford: Basil Blackwell.

Census of India 2001, *Provisional Figures at a Glance for India and States/UTS: Census of India 2001*.

Dagar, J.C., A.D. Mongia and A.K. Bandopadhyay (1991), *Mangroves of Andaman and Nicobar Islands*, New Delhi: Oxford & IBH Publ. Co.

Durrenberger, E.P. (2001), 'Anthropology and Globalization', *American Anthropologist*, **103**(2), 531–5.

EEA (2003), *Europe's Environment: The Third Assessment*, Copenhagen: European Environment Agency (EEA), Environmental Assessment Report No. 10.

Fischer-Kowalski, Marina (1997), 'Society's Metabolism: On Childhood and Adolescence of a Rising Conceptual Star', in Michael Redclift and Graham Woodgate (eds), *The International Handbook of Environmental Sociology*, Cheltenham, UK and Lyme, US: Edward Elgar, pp. 119–37.

Fischer-Kowalski, M. and Helmut Haberl (1993), 'Metabolism and Colonization. Modes of Production and the Physical Exchange between Societies and Nature', *Innovation: the European Journal of Social Science Research*, **6**, 415–42.

Fischer-Kowalski, M. and Karl-Heinz Erb (2003), 'Gesellschaftlicher, Stoffwechsel in Raum. Auf der Suche nach einem sozialwissenschaftlichen Zugang zur biophysischen Realität', in Peter Meusburger and Thomas Schwan (eds), *Humanökologie. Ansätze zur Überwindung der Natur-Kultur-Dichotomie*, Stuttgart: Steiner Verlag, pp. 257–85.

Fischer-Kowalski, M., F. Krausmann and B. Smetschka (2004), Modelling Scenarios of Transport Across History from a Socio-metabolic Perspective', *Review*, Fernand Braudel Center, **27**, 307–42.

Fukui, Hayao (1993), *Food and Population in a Northeast Thai Village*, Honolulu: University of Hawaii Press.

Graeber, David (2002), 'The Anthropology of Globalization (with Notes on Neomedievalism, and the End of the Chinese Model of the Nation-State)', *American Anthropologist*, **104**(4), 1222–7.

Grünbühel, Clemens M., Helmut Haberl, Heinz Schandl and Verena Winiwarter (2003), 'Socioeconomic Metabolism and Colonization of Natural Processes in SangSaeng Village: Material and Energy Flows, Land Use, and Cultural Change in Northeast Thailand', *Human Ecology*, **31**(1), 53–87.

Grünbühel, Clemens M. and Heinz Schandl (2005), 'Using Land-Time-Budgets to Analyse Farming Systems and Poverty Alleviation Policies in the Lao PDR', *International Journal of Global Environment Issues*, **5**(3/4), 142–80.

Haberl, Helmut (2001a), 'The Energetic Metabolism of Societies, Part I: Accounting Concepts', *Journal of Industrial Ecology*, **5**(1), 11–33.

Haberl, Helmut (2001b), 'The Energetic Metabolism of Societies, Part II: Empirical Examples', *Journal of Industrial Ecology*, **5**(2), 71–88.

Haberl, Helmut, Karl-Heinz Erb, Christof Plutzar, Marina Fischer-Kowalski and Fridolin Krausmann (2007), 'Human Appropriation of Net Primary Production (HANPP) as Indicator for Pressures on Biodiversity', *Ecological Questions*, in print.

Hanks, Lucien M. (1972), *Rice and Man. Agricultural Ecology in Southwest Asia*, Honolulu: University of Hawaii Press.

Kelly, R.L. (1995), *The Foraging Spectrum: Diversity in Hunter-Gatherer Lifeways*, Washington: Smithsonian Institution Press.

Keyes, Charles F. (1967), *Isan: Regionalism in Northeastern Thailand*, Southeast Asia Program, Department of Asian Studies, Ithaca, NY: Cornell University.

KKU-Ford Cropping Systems Project (1982), *An Agroecosystem Analysis of Northeast Thailand*, Khon Kaen: Faculty of Agriculture, Khon Kaen University.

Klausner, William J. (1993), *Reflections on Thai Culture. Collected Writings of William J. Klausner*, London: Siam Society.

Krausmann, Fridolin, Helmut Haberl, Karl-Heinz Erb and Mathis Wackernagel (2004), 'Resource Flows and Land Use in Austria 1950–2000: Using the MEFA Framework to Monitor Society–Nature Interaction for Sustainability', *Land Use Policy*, **21**(3), 215–30.

Man, E.H. (1903), 'Report on the Penal Settlement in Nancowry Harbour [originally appeared as *Annual Report 1888–89*]', in Anonymous (eds), *Richard C. Temple. The Andaman and Nicobar Islands: Report on the Census*, Calcutta: Government of India Central Printing Office.

Mayrhofer-Grünbühel, Clemens (2004), 'Resource Use Systems of Rural Smallholders. An Analysis of Two Lao Communities', Dissertation, University of Vienna, Vienna.

Meleisea, Malama (1980), *O Tama Uli: Melanesians in Western Samoa*, Suva: Institute of Pacific Studies, University of the South Pacific.

Müller-Herold, Ulrich and Rolf P. Sieferle (1998), 'Surplus and Survival: Risk, Ruin and Luxury in the Evolution of Early Forms of Subsistence', *Advances in Human Ecology*, **6**, 201–20.

Polanyi, Karl (1957), *The Great Transformation: The Political and Economic Origins of Our Time*, Boston, MA: Beacon Press.

Ramitanondh, Shalardchai (1989), 'Forests and Deforestation in Thailand: A Pandisciplinary Approach', in Siam Society (eds), *Culture and Environment in Thailand*, Bangkok: Duang Kamol, pp. 23–50.

Sahlins, Marshall (1972), *Stone Age Economics*, New York: Aldine de Gruyter.

Sakamoto, S. (1996), 'Glutinous-Endosperm Starch Food Culture Specific to Eastern and Southeastern Asia', in Roy Ellen and Katsuyoshi Fukui (eds), *Redefining Nature. Ecology, Culture and Domestication*, Oxford, Washington, DC: Berg, pp. 13–26.

Saldhana, C.J. (1989): *Andaman, Nicobar and Lakshadweep: An Environmental Impact Assessment*, New Delhi: Oxford & IBH Publ. Co.

Sankaran, R. (1998), *The Impact of Nest Collection on the Edible-nest Swiftlet Collocalia Fuciphaga in the Andaman and Nicobar Islands*, Coimbatore: Sálim Ali Centre for Ornithology and Natural History.

Schandl, Heinz, Clemens M. Grünbühel, Helmut Haberl and Helga Weisz (2002), *Handbook of Physical Accounting. Measuring Bio-physical Dimensions of Socio-economic Activities. MFA – EFA – HANPP*, Vienna: Federal Ministry of Agriculture and Forestry, Environment and Water Management.

Sieferle, Rolf P. (1997), *Rückblick auf die Natur: Eine Geschichte des Menschen und seiner Umwelt*, Munich: Luchterhand.

Singh, Simron J. (2003), *In the Sea of Influence: A World System Perspective of the Nicobar Islands*, Lund: Lund University.

Singh, Simron J. and Clemens M. Grünbühel (2003), 'Environmental Relations and Biophysical Transitions: The Case of Trinket Islands', *Geografiska Annaler, Series B, Human Geography*, **85 B**(4), 187–204.

Singh, Simron J., Clemens M. Grünbühel, Heinz Schandl and Niels B. Schulz (2001), 'Social Metabolism and Labour in a Local Context: Changing Environmental Relations on Trinket Island', *Population and Environment*, **23**(1), 71–104.

Temple, Richard C. (1903), *The Andaman and Nicobar Islands. Report on the Census*, Calcutta: Government of India Central Printing Office.

Thongmanivong, Sithong (2004), 'Sustainable Use of Land and Natural Resources in Lao PDR. Material Energy Flow Accounting, Human Appropriation of Net Primary Production', Dissertation, University of Vienna: Vienna.

Weisz, Helga, Marina Fischer-Kowalski, Clemens M. Grünbühel, Helmut Haberl, Heinz Schandl and Verena Winiwarter (1999), *Sustainability Problems and Historical Transitions – A Description in Terms of Changes in Metabolism and Colonization Strategies*, Proceedings of the International Conference, Nature, Society and History: Long-term Dynamics of Social Metabolism, 30 September – 2 October 1999, Vienna, Austria (CD-ROM).

Wongthes, Pranee and Wongthes Sujit (1989), 'Art, Culture and Environment of Thai-Lao Speaking Groups', in Siam Society (eds), *Culture and Environment in Thailand*, Bangkok: Duang Kamol, pp. 161–9.

7. Transition in a contemporary context: patterns of development in a globalizing world

Nina Eisenmenger, Jesus Ramos Martin and Heinz Schandl

7.1 INTRODUCTION

Three earlier chapters in this book discuss the process of industrialization in Europe, taking the United Kingdom and Austria as examples for early and late transition respectively from an agrarian, solar-based socioecological regime to a fossil-fuel-based industrial regime. These are examples of endogenously driven industrialization in which domestic agricultural development yielded a surplus that could be invested in other sectors of the economy. Recent transitions in countries, often denoted as developing countries, are confronted with quite a different development context. They are integrated in a global economic system where the division of labour assigns each country a specific role within the world economy. Thus, socioeconomic change within a developing country is not mainly dependent on internal processes, but rather is strongly influenced by the international context. This has become particularly evident since the 1980s, the time period for this analysis, during which the accelerated incorporation of developing economies into world markets occurred.

This chapter focuses on two world regions, South America and Southeast Asia, and takes a comparative view. The analysis is based on a set of countries in these regions, that is, Lao PDR (Laos), Vietnam, Philippines and Thailand in Southeast Asia, and Venezuela, Brazil and Chile in South America, for all of which we engaged in detailed case studies except for Chile (Giljum 2004). Both South America and Southeast Asia are developing regions undergoing a rapid transition process. They are very different in terms of their socioeconomic characteristics. We describe biophysical aspects of the economies of selected countries in these regions in order to analyse different paths of transition in socioeconomic development and resource use. We begin the analysis in South America and Southeast Asia by focusing on the agricultural sector, where we see different structures and different dynamics in existence. A considerable proportion of agricultural production in South America was taken over relatively early on by large companies, with United Fruit Company arriving in Guatemala in 1901 (Dosal 1993), and the rural population faced the choice of either being employed in agricultural wage labour or moving to urban centres to make a living there. In Southeast Asia, on the other hand, small-scale farming remains predominant. Many smallholders produce for subsistence outside the market, which limits capitalization of the agricultural holdings. However, in Southeast Asia, cash cropping by large companies exists in parallel to subsistence production by smallholders.

In both South America and Southeast Asia, for different reasons, agricultural production did not generate a sufficient amount of economic value-added to develop industrial production and manufacturing. Instead, international capital from foreign direct investment and loans supported the development of the second and third sectors from the 1960s onwards. In South America, the failure of the imports substitution strategy in the 1960s (Bruton 1998) resulted in a strategic reorientation towards export-led development, supported by international loans. In the case of Southeast Asia, significant foreign direct investment has been particularly evident since the late 1970s and early 1980s (Ritchie 2002, p. 12). Loans promoted by the World Bank and the International Monetary Fund (IMF) are bound to structural adjustment programmes. These programmes and the foreign investment are aimed primarily at intensifying integration into the world market and at enlarging production for exports. This goes hand in hand with economic liberalization characterized by weak labour regulation, the lifting of trade tariffs and opening local markets to foreign capital and goods. The exports a country specializes in depend on the available resources and on which competences the country has accumulated in the course of its economic history. South America has a long history of providing raw materials to the United States and other industrial centres and specializing in the intensive extraction of natural resources for exports. We will show how this strategy results in a typical metabolic profile. Southeast Asia has taken a different path of development, since its endowment with natural resources such as ores and industrial minerals is relatively small. On the other hand, Southeast Asia is densely populated and can supply an abundant labour force. Transition in Southeast Asia has therefore moved towards specializing in labour-intensive light industrial production (for the environmental problems related to the rapid industry-led growth, see Rock, Angel and Feridhanusetyawan 1999). A comparison between the labour and energy intensity of the two regions confirms the existence of these different patterns.

These two regions, South America and Southeast Asia, show that recent transitions in developing countries can follow different trajectories that can be detected using classical social-science-based indicators (for example, GDP) or socioecological or biophysical ones. The different trajectories also imply a different impact upon the environment, as measured by material and energy flows. Each of the trajectories is marked by constraints. In both cases an abundant labour force exists, used mainly in agriculture in Southeast Asia, and in services in South America. This excess labour force may eventually shift to industry, thereby introducing the opportunity for an economic strategy based on low-cost labour. However, the two regions have to escape from this 'specialization trap' (Ekins, Folke and Costanza 1994; Muradian and Martinez-Alier 2001) if they want to increase their economic productivity to cope with population growth and the rise in material standards of living.

7.2 ECONOMIC DEVELOPMENT AND GLOBAL INTEGRATION OF SOUTH AMERICA AND SOUTHEAST ASIA

As has been shown in the earlier chapters of this volume, historical transitions in countries that industrialized comparatively early were mainly based on changes of internal patterns and processes. The agricultural sector was the motor and starting point for the transition process, because of increases in labour and land productivity. A modernization process gained ground, which created surplus that could be used to fuel other sectors of the economy and to create the necessary infrastructure. Finally, subsistence production ceased and the agricultural sector was transformed into a market-oriented and, since the 1950s, highly industrialized sector with high land and labour efficiencies. Within industrial economies today, the agricultural sector's contribution to GDP as well as to employment is below 5 per cent. However, in physical terms, the amount of agricultural production in many industrialized economies is greater than ever before.

Contemporary transition processes in developing countries take place under very different conditions. Integration into world markets and high international mobility of resources, goods and capital significantly influence the development of these transitional economies. This different context provides new options for, but also constraints upon, development compared with the transition of early industrialized countries. Before we turn to biophysical patterns of transitions in developing regions, we have to take into account international economic processes in which South America and Southeast Asia have been and are involved.

In economic theory, the basis for international trade is the existence of interregional differences concerning endowment with natural resources (and therefore raw materials), technology and climatic conditions. According to this approach, trade widens the growth potential of nations by making resources available that are not locally based and making produce marketable for which local demand would be too low. Both Smith's (1776) classic theory on international trade based on absolute advantages and Ricardo's (1817) theory based on comparative advantages, dealt with this issue long ago and were complemented in the early 20th century by the 'staple theory of growth'[1] (Innis 1930) and the Heckscher-Ohlin trade model[2] (Heckscher 1919; Ohlin 1933).

Trade, according to the classical understanding, is leading us to a situation in which all economies would finally gain advantages. During the 1950s and 1960s, economists were already discerning diverging national pathways to industrialization, even where countries departed from similar starting points. New theories evolved, aiming to explain why this should be the case. These drew upon new applications of theoretical concepts of 'imperialism' (revived by Paul Baran 1957, based on Rosa Luxemburg [1913] 2003 and Lenin [1917] 1971, and with major contributions from Mandel 1968), the 'dependency theory' (developed by the Argentinean economist Raul Prebish 1950, 1959, with later contributions from the former Brazilian president Fernando Henrique Cardoso [Cardoso and Faletto 1979]), and the 'world systems perspective' (Amin 1976; Emmanuel 1972; Frank 1967; Wallerstein 1979). The common ground of all three approaches lay in the notion that in the existing world system, peripheral countries specialize in the production of primary commodities such as raw minerals and agricultural products that are less technologically sophisticated, are more labour-intensive and exposed to severe competition on world markets, thus leading to low prices and low surplus. Primary products are then exported to industrialized cores. These are characterized by a high level of capital accumulation and complex production activities. Production is based on advanced technologies, highly mechanized production structures and higher wages (Shannon 1996). Industrial cores sell their high-tech and capital-intensive products to the peripheral countries. This exchange on world markets is leading to an outflow of surplus from the periphery to the core based on two processes. First, peripheral countries specialize in exports of agricultural products and raw materials, where they are confronted with increasing competition from other developing countries, which forces them to reduce prices to keep export revenues. This leads to a worsening in the terms of trade, and peripheral countries have to export ever more goods in order to obtain the same revenues to support the imports needed (machinery is produced with a more oligopolistic structure and therefore higher prices can be gained). Second, low salaries are found in the periphery due to the massive 'reserve army' of labour force generated through technological progress in agriculture. Revenues from increased efficiency thus result in lower prices of exports instead of increased income for workers (Emmanuel 1972).

Economic development in the periphery is therefore complementary to economic development in the centre. Specialization in exports of raw materials, in the medium and long run, is supporting underdevelopment in the periphery and development in the centre. The situation for developing countries appears even worse when we consider that specializing in exporting raw materials leads to a depletion of domestic natural resources by selling out the domestic resource base.

International financial support for developing countries is given by the World Bank and the IMF. As mentioned above, loans are bound to structural adjustment programmes, now called poverty reduction strategies, that focus on liberalization of markets, by opening both financial and commodities markets to foreign products. The latter often occurs before the infant industry within the developing country has achieved a minimum size necessary for competing at the global level (Stiglitz 2002). At the same time, protectionism in developed economies prevents developing countries from benefiting from their comparative advantage in the production of labour-intensive goods. This explains why some developing countries still focus on producing resource-intensive goods, even though competition among developing countries is driving prices down.

This problem seems to be recurrent, since prices for raw materials and commodities from labour-intensive production are declining as they did during most of the 20th century (Schor 2005), with a worsening of the terms of trade (see Bloch and Sapsford 1997; Prebish 1950; Sapsford and Balasubramanyam 1994; Singer 1950; Stiglitz 2002; Zanias 2005). According to Bloch and Sapsford (1997), terms of trade of primary producers declined by more than 1 per cent per annum in the period 1948–86.[3] Zanias (2005), on the other hand, found that the worsening of the terms of trade between 1900–98 was not smooth, but, on the contrary, developed stepwise, with two major breaks in the years 1920 and 1984. The cumulative effect was a drop of 62 per cent in the terms of trade, to a level of one-third of that at the beginning of the 20th century. Of course this fall has huge implications for developing countries specializing in exporting primary commodities, since it may render their specialization detrimental in the long run. Moreover, the stepwise character of this deterioration makes it more difficult for those countries to anticipate such shocks, rendering their economies much more fragile where there is no diversification of exports.

A global division of labour is not detrimental in itself, if it leads to a process of surplus generation, and therefore capital accumulation in both trading partners that will drive the change towards industrialization. Porter (1990) introduced the 'stages theory of competitive development', which was first envisaged for the study of competitiveness at the company level but may also give insights at the national level. Competitive advantage develops according to four stages: 1) factor-driven; 2) investment-driven; 3) innovation-driven; and 4) wealth-driven. Most developing countries may still be classified as being at the factor-driven stage, which relies on natural resource-based activities or labour-intensive manufacturing, while a few are entering the investment-driven stage. According to Ozawa (2006), the factor-driven stage is related to factor-trade advantages (that is, inherent in the production factor itself and translating into lower prices of goods and commodities), whereas the investment stage relies on scale-based advantages (that is, related more to the production process itself and less to the production factors).

As we can see, developing economies are confronted with the challenge of moving from the factor-driven stage to the investment-driven stage but in order to do so, they need to increase the level of capital accumulation through investments in infrastructure and machinery that allow further increases in economic productivity. At present, some of these countries are failing to achieve this because: 1) their population is growing rapidly, so most available resources are required to provide this population with their basic needs; 2) they have to pay the debt service from the most recent regional financial crisis; 3) the prices of the goods they export are in decline.

7.3 TRANSITION IN THE AGRICULTURAL SECTOR

During the Industrial Revolution, surplus from the agricultural sector provided support for economic development in other sectors. Land and labour productivity increased and more people, engaged in sectors other than agriculture, could be fed with domestic products from agriculture. Urban centres emerged that depended on rural agricultural production. This process of agricultural modernization relying on traditional resources was made possible by a combination of institutional changes and changes in the mode of production and has been the subject of in-depth studies by numerous authors. The agricultural sector in parts of Europe and the United States has only been integrated in the industrial regime since the 1950s, when fossil fuels, chemical fertilizers and pesticides began to boost agricultural production. Animal labour vanished and land and labour productivity rose to unprecedented levels.

Agriculture still plays a crucial role in the contemporary economic life of developing countries. In Southeast Asia, around 50 per cent of the labour force is occupied in agriculture. Agriculture provides about 13 per cent of GDP.[4] In most countries in Southeast Asia, people living in rural areas account for more than 60 per cent of the total population. South America, in contrast, is a region with huge urban centres and 80 per cent or more of the total population living in urban settings. Even at an early stage, large parts of the agricultural production in South America were taken over by large companies and rural populations moved to urban centres (Bairoch 1975). Agriculture occupies 18 per cent of the labour force and contributes 8 per cent to GDP (FAO 2004).[5]

In interpreting these averages, we should be aware that the history of colonization established a dual system of agriculture in most developing regions. Crop production for export purposes was usually dominated by plantation systems and was based on advanced technologies financed by foreign investment. Crops produced in plantations include bananas, cocoa, coffee, tea, sugar cane, natural rubber, cotton fibre, jute and oil seeds. Subsistence agriculture, on the other hand, was often driven away from the most fertile land and almost always used traditional low-input methods.

Taking the ratio between agricultural and non-agricultural workers as a proxy for agricultural labour productivity (Overton and Campbell 1999), and thus for the involvement of capital in the agricultural sector, we can identify South America as being more advanced in agricultural labour productivity, since one worker in agriculture supports five to ten workers in other sectors. In Southeast Asia, the ratio is one to one, which reflects the lack of capital, infrastructure and technology in the region. Since 1960, labour productivity in Southeast Asian agriculture has remained static, whereas South America has experienced rapid improvements in labour productivity through the application of machinery, fossil fuel energy and chemicals.

Differences in the proportion of land utilized for agricultural production provide a similar picture of two different patterns. Southeast Asia uses 28 per cent of its land area as cropland with a significant increase over the last 40 years (this figure was 18 per cent in 1961). In South America, on the other hand, only 7 per cent of total land area is used as cropland (compared with 4 per cent in 1961), but 23 per cent of its land area is used as permanent pastures, a land-use category that is negligible in Southeast Asia (only 3 per cent in 2000) (FAO 2004).

These characteristics of land use in agriculture and for livestock husbandry in the two regions are closely related to population density. In Southeast Asia, high population density and a large share of agricultural labour force result in low land availability per agricultural worker. A common average figure for land availability in Southeast Asia in 2000 as well as in 1960 was 0.6 hectares (ha) of land per labourer, and land may have become even more scarce since 2000. For example, 0.3 ha per worker have to suffice as the basis of production in Vietnam. In contrast, one South American labourer had an average of 4.7 ha at his or her disposal in 2000, compared with 2.3 hectares in 1960 (FAO 2004). The labour-intensive agricultural production in Southeast Asia only yields an economic output of less than 1000 USD[6] per worker per year (UN Statistics Division 2004). This figure has not increased significantly during the last 40 years. Labour productivity in South America is much higher and has increased impressively since 1960. Starting at around 1500 USD/worker in 1960, output rose to 3500 USD/worker in the year 2000 (UN Statistics Division 2004), a yearly growth rate of 21 per cent. Even this much higher value, however, is easily surpassed by the high-performance agriculture in the United States. Agricultural value-added per worker in the United States was around 20 000 USD/worker in 1960 and reached 50 000 USD/worker in 2000.

Even if Southeast Asian agriculture may appear backward, operating with low inputs and partly outside the market, it serves important social functions. For example, it is a pool of labour force that can be moved to other sectors if necessary but can also absorb labour force in times of economic crisis. This is not the case in South America where most of the 'excessive' rural labour force has already migrated to cities and can be found in the low value-added services sector.

Economic land productivity shows a reverse picture, with very high value-added per land area for some Southeast Asian countries, for example, 1000 USD/ha/yr in the Philippines and 700 USD/ha/yr in Thailand, resulting from the high intensity of use. Even Vietnam and Laos both achieve a land productivity rate of 450 USD/ha/yr, which is much higher than that in South America, where the average is only 100–250 USD/ha/yr. In contrast, physical yields (kg/ha/yr) are higher in South America both for staples and for cash crops.[7] However, yields in South America are often achieved through high inputs, while reasonable yields in Southeast Asia are mostly a result of intensive labour inputs and multi-cropping, that is, more than one harvest per year.

Another indicator of agricultural industrialization is the use of chemical fertilizers. Interestingly, fertilizer consumption per hectare of arable land and permanent crops is around 80 kg/yr in most of the Southeast Asian and South American countries (FAO 2004). This is similar to fertilizer consumption in the United States. Exceptions are Laos with negligible fertilizer input, Vietnam with 250 kg/ha/yr and Chile with 200 kg/ha/yr. In Vietnam and Chile, fertilizer use is similar to that in Europe and Japan, both countries or regions with intensive agriculture. Data on fertilizer consumption over time reveal that South American countries experienced a sharp increase in the 1970s while Southeast Asia did not follow this path until the 1980s. Industrialization and mechanization are also reflected in the use of machinery such as tractors for agricultural production (FAO 2004). Here again, figures do not differ greatly. Most countries have around 12 tractors per 1000 ha of arable land or permanent crops at their disposal. Again, Vietnam and Chile use more machinery, that is, 20 tractors/ha. Similar to the fertilizer consumption described above, Southeast Asian countries first show a significant increase in the 1990s, whereas in South America, machinery use had already surged during the 1970s.

To conclude, we can identify similarities but also differences between the agricultural sectors in South America and Southeast Asia. Both regions are large producers of agricultural goods, which contribute to their respective balance of payments. However, agriculture in South America is more industrialized, on average, being based on higher inputs and being less labour-intensive. In South America, land is more abundant than in Southeast Asia and its use is characterized by intensification strategies and landownership structures of larger holdings that allow for a concentration of capital and a greater disposition to invest in agricultural activities. In many cases, landless people either have to labour for a salary or move to the urban centres in order to make a living.

Southeast Asia is to a larger degree characterized by rural landownership and subsistence production by smallholders. Capital is often lacking and therefore inputs are low. Reasonable production levels are often achieved through high workload of rural households to achieve food security and to gain income by selling surplus production on the market. This is not to say that Southeast Asian countries do not produce for the world market. Especially concerning rice, Vietnam and Thailand are large producers on the global scale. In contrast to the countries involved in the first wave of industrialization, the agricultural sectors in South America and Southeast Asia do not provide large investments to the secondary and tertiary sectors of their domestic economies.

Patterns of Resource Extraction and Trade

Parallel to the industrialization of the agricultural sector, relevant parts of these economies were export-oriented from a relatively early stage, including extractive activities such as crude oil production and mining of ores, as well as traditional industrial branches such as textile manufacturing and innovative sectors such as electronics. This transition to industrialized production with particular specializations and integration into the global

188   *Socioecological transitions and global change*

division of labour led to different metabolic patterns, and we will now turn to an analysis of the material flows and the use of natural resources in the two regions.

Material flow accounting (MFA) is an accounting tool that traces all material flows into and out of the economic system and provides different aggregated indicators for resource use such as domestic extraction (DE), domestic material consumption (DMC), and physical trade balance (PTB)[8] (Eurostat 2001). Material flows are reported in different material groups, most commonly biomass, fossil fuels and minerals, or even in more detail (Weisz et al. 2006). In this section, we will compare MFA results of the two regions to three OECD countries or country aggregates, namely the United States, Japan (Adriaanse et al. 1997; Matthews et al. 2000) and the EU-15 (Weisz et al. 2006).[9] In our comparison of resource use, we draw on the last available year for the material flows.

Domestic extraction per capita in Southeast Asian countries is 4 tons per capita and year (t/cap/yr) in Laos, 7.4 t/cap/yr in the Philippines and around 10 t/cap/yr in Thailand and Vietnam. Biomass extraction per capita is highest in Thailand (around 3 t/cap/yr) and Laos (2.3 t/cap/yr) and rather low in the Philippines and Vietnam (1.7 and 1.4 t/cap/yr respectively). In the case of Laos, biomass accounts for 70 per cent of total domestic extraction and between 14 per cent and 28 per cent in the other countries. Extraction of fossil fuels occurs in Thailand and Vietnam while amounts extracted in Laos and the Philippines are negligible.[10] Primary production of industrial minerals is significant in Vietnam, Thailand and the Philippines, whereas Laos only extracts minor amounts. Economic development seems to go hand in hand with a significant build-up of infrastructure, reflected in the large amounts of construction minerals consumed in all Southeast Asian countries (with the exception of Laos).

The South American countries studied here showed considerably higher domestic extraction per capita than Southeast Asia. Domestic extraction per capita of Venezuela and Brazil was 15 t/cap/yr. Fossil fuels constitute the main share of domestic extraction in Venezuela (around 9 t/cap)/yr, whereas Brazil extracts large amounts of biomass (9 t/cap/yr). Chile has a very high domestic extraction amounting to 40 t/cap/yr. This value is mostly due to mining of copper ores for export. Chilean copper mining is characterized by its low metal content of 1 per cent (see United States Bureau of Mines 1987). Extraction of copper thus produces high amounts of ancillary mass, which by definition is included in MFA (Eurostat 2001). For comparison, the domestic extraction per capita in OECD countries ranges between 10 t/cap/yr in Japan and 21 t/cap/yr in the United States; EU-15 countries extract on average 13 t/cap/yr (Figure 7.1).

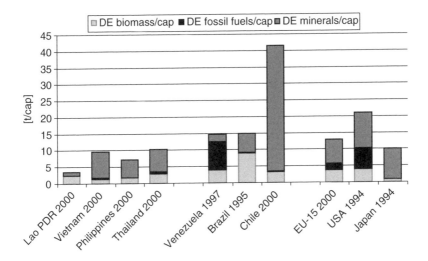

Figure 7.1   Domestic extraction (DE) per capita

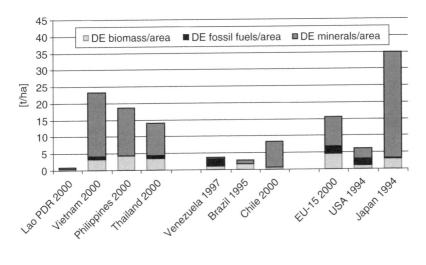

Figure 7.2   Domestic extraction (DE) per land area

Countries with high population densities are typically characterized by a low domestic extraction per capita, but domestic extraction is high per unit of land area (Figure 7.2) of their territory (Weisz et al. 2005). Of all countries studied here, Japan, which has a high population density of 330 persons per km², has the highest DE per land area in our sample at 35 t/ha/yr. The Philippines and Vietnam follow with population densities of 250 and 235 persons per km² and corresponding DE per land area of 18 and 23 t/ha/yr respectively. Thailand and the EU-15 countries have population densities of about 120 persons per km² and extract around 15 t/ha/yr. In comparison, the other countries with population densities of around 20 to 30 persons per km² only extract 3–6 t/ha/yr. Laos has an even lower DE with around 1 t/ha/yr, whereas Chile's DE is much higher at 8 t/ha/yr, due to its large mining sector.

Summarizing these results, we can see that countries in Southeast Asia extract less material per capita than countries in South America. The EU-15 countries and Japan fall between these extremes. Chile and the United States do not fit the general pattern and have an unusually high domestic extraction per capita.

In terms of material fractions, Southeast Asian countries extract comparably low amounts of biomass per capita, but the amount of biomass extracted per land area shows highest values among the assessed countries. Considering the fact that Southeast Asian countries all have very high population densities, agricultural intensification seems to be common, suggesting that high environmental pressures on land exist. Comparable figures for biomass extraction per land area are only achieved in the EU-15 countries. Here, population density is high, as is biomass extraction per capita. This is a result of the agricultural policy of the European Union, which protects domestic markets through state subsidies. Japan's population density is much higher than that of the EU-15, resulting in low biomass DE per capita. However, if expressed as per unit of land available, Japan's biomass DE is still high.

Fossil fuels are a resource strongly linked to national availability. Venezuela, the United States and the EU-15 countries have fossil fuel deposits, resulting in high amounts of domestic extraction. The other countries show no significant domestic extraction of fossil fuels with the exception of Vietnam, which has a potential for future growth.

Minerals, the last category discussed here, are composed of construction minerals, industrial minerals and ores. The first group of construction minerals includes bulk material flows of low economic value that are mainly used locally. Industrial minerals and ores, on the other hand, are resources for industrial production that are frequently traded on international markets. Due to the varying quality of data it was not possible, however, to separate these two groups of minerals in data for all countries. An interpretation of the results available is therefore very difficult. What may be said with some certainty is that economic growth goes hand in hand with large amounts of construction materials for infrastructure provision (Weisz, Krausmann and Sangarkan 2006). This is also reflected in economic data: the Association of Southeast Asian Nations, ASEAN,[11] reports World Bank estimates that developing Southeast Asian countries would need between 1.2 trillion USD and 1.5 trillion USD in investment in infrastructure until 2010 to enable infrastructure to keep pace with economic growth. Extraction of industrial minerals and ores on the other hand is dependent on national deposits. According to mineral statistics (Crowson 2001), South American countries have larger reserves of minerals than those in Southeast Asia and therefore extract more minerals from their domestic environment.

Domestic material consumption (DMC) (Figure 7.3) is a measure for the material resources used in an economy either in production sectors (intermediate consumption) or in final consumption. In OECD countries, on average, minerals make up 50 per cent of DMC (most of this being construction minerals), fossil fuels constitute 30 per cent and biomass 20 per cent. The situation is obviously different in developing countries, where biomass plays a more important role and the use of fossil fuels is often negligible.

Biomass use is of higher relative importance in South America, where in two of our case studies, Venezuela and Brazil, biomass use constitutes about half of overall DMC. This is due to export crops and a large livestock sector. The relative importance of biomass resources in Southeast Asian economies is much lower, its use constituting around 15 per cent of DMC in Vietnam and around 25 per cent of DMC in the Philippines and Thailand. Laos is one of the least developed countries in Southeast Asia with an economy that is primarily based on agriculture, resulting in a biomass consumption of over 60 per cent of DMC.

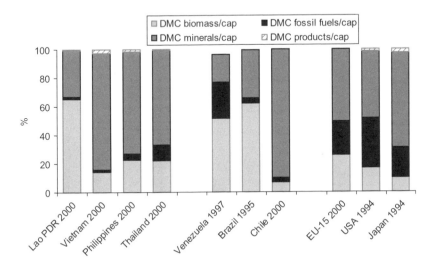

Figure 7.3   *Share of biomass, fossil fuels, minerals and products in*
            domestic material consumption (DMC), in percentage terms

The share of fossil fuel use is rather low in all developing countries with the exception of Venezuela, which is a large producer of crude oil and subsidizes oil products for domestic consumption. This results in domestic consumption of fossil fuels at OECD levels in Venezuela. Chile is a special case with an exceptionally high share of mineral use due to the large copper exports.

In the year 2000, consumption of material resources per capita was higher in OECD countries (between 15 and 20 t/cap) than in developing countries, with the exception of Chile. Per capita consumption of materials in Chile is around 40 t/cap/yr, due to the huge volumes of gross ore containing copper. Brazil has a domestic consumption of 15 t/cap/yr, which is similar to the EU-15 and Japan. Venezuela's resource consumption is lower and closer to the Southeast Asian pattern. On average, domestic consumption is lower in Southeast Asian countries. While Thailand and Vietnam consume around 10 t/cap/yr, the corresponding figure is around 8 t/cap/yr for the Philippines. Laos has a per capita consumption of material resources of below 5 t/cap/yr, which reflects the status of the country as a least-developed country (Figure 7.4). Only Laos, the Philippines and Venezuela fall below the global average for domestic material consumption of 8 t/cap/yr (Schandl and Eisenmenger 2006). Clearly, the world average is greatly influenced by large countries such as India, China and Indonesia, which have low rates of material consumption. This may be expected to change as a consequence of their pronounced growth.[12]

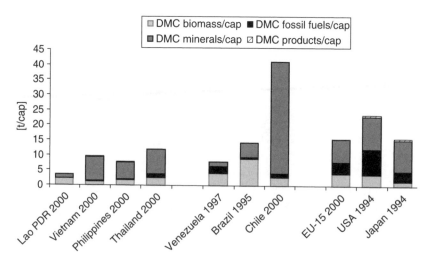

Figure 7.4   Domestic material consumption (DMC) per capita

It has been argued that per capita values of domestic material consumption not only reflect differences in production patterns and the standard of living, but are a function of population density (Weisz et al. 2006) to a certain extent. As is the case for DE, densely populated countries have lower DMC/capita while less populated countries have higher per capita DMC, while the reverse picture evolves when projected onto land area. The world average of DMC in the year 2000 was 8 t/cap and 2.83 t/ha (Schandl and Eisenmenger 2006) (Figure 7.5).

The transition from an agrarian solar-based economy to a fossil-fuel-based economy also changes the patterns of material and energy use. This leads first of all to an overall increase of materials extracted and used. In empirical MFA data we may discern this by comparing Laos, a country that can still be considered as primarily agrarian, with the other countries of our sample. With the overall increase in materials used we can observe a shift from a metabolic pattern that is mainly biomass-based to one that makes increasing use of mineral materials. A high rate of biomass use is only maintained by countries where animal husbandry is an important economic factor. The same shift becomes visible in energy data, where the share of energy from biomass is reduced and energy from fossil fuels increasingly gains importance.

As already mentioned, another indicator for transition could be the amount of construction minerals used. Transition from an agrarian to an industrialized country goes hand in hand with an expansion in built infrastructure such as buildings and roads, which is directly linked to the amount of construction minerals used.

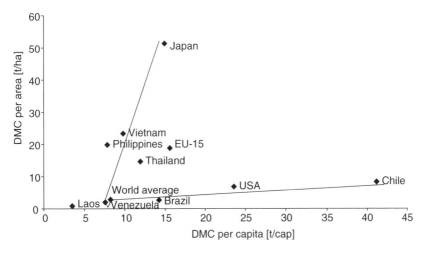

Figure 7.5   Domestic material consumption per capita per area

Socioecological transitions and global change

Integration into the World Economy – Biophysical Trade Patterns

In a global economy, a specific division of labour is established and materials and commodities are traded between the countries involved. The materials used within an economy are therefore a mixture of domestic resources and imported materials, minus those materials exported. We use physical trade balances (PTB) (Figure 7.6), which are calculated by subtracting the mass of material exports from imports (Eurostat 2001), to assess biophysical trade patterns. A positive PTB value represents net imports of materials in terms of weight, whereas negative values express net exports.

Aggregated figures of per capita PTB suggest an unambiguous picture of regional patterns: all Southeast Asian countries are net importers of materials, whereas South American countries are exporters. Both Venezuela and Brazil are net exporters of primary resources. In the case of Venezuela, this is mainly crude oil (more than 6 t/cap/yr) and in the case of Brazil, consists mainly of metal ores such as iron ore. Chile has an even trade balance due to its considerable import of fossil fuels, mainly gasoline, diesel oil and gas from Argentina, but it is a net exporter of both biomass and metal ores, mainly copper. In the case of Chile, it has to be considered that gross copper ores include huge amounts of ancillary mass that is extracted but not further processed and not included in the traded materials moved across the border. Most of this material is separated from copper during the first phase of processing and is returned to the natural environment as waste. These waste materials from copper mining stay within the Chilean environment, whereas

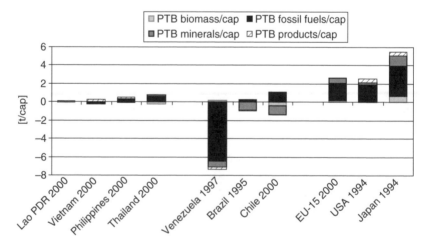

Figure 7.6   Physical trade balance (PTB) per capita per year, by material category

concentrated copper is exported. If we were to consider waste materials from economic processing as being related to exports and to allocate these 'indirect flows' (Eurostat 2001) to the corresponding exports, Chile's material outflows would obviously be much higher.

As expected, the selected OECD countries are net importers of natural resources. Fossil fuels constitute a major part of these imports, but imports of minerals, mainly industrial minerals and ores, also play an important role, even more so when considering the developments of the last 30 years where we see that these materials are increasingly traded and substitute domestic extraction and production (Weisz et al. 2005). In the case of the EU-15, and even more so for Japan, limitations in land availability also result in substantial imports of biomass.

As already anticipated in the section on domestic extraction, trade figures now complement the picture of South American countries as suppliers of raw materials to world markets. Southeast Asian countries, on the other hand, are limited in space so that resource extraction for bulk exports of raw materials is neither possible nor supported by historical development. The claim that developing countries exploit their natural resources for exports would only hold true for South American countries. The expression 'extractive economies' (Bunker 1985) characterizes these countries well.

Dividing monetary volumes of trade flows by material imports and exports measured in tons results in an average monetary value of 1 ton of traded material or good. During the lifecycle of a product, monetary value rises with each phase of economic processing, whereas material content decreases (Fischer-Kowalski and Amann 2001). Thus, raw materials typically have a large mass but little economic value. By contrast, finished products mostly contain much smaller amounts of materials but have a higher monetary value. Figure 7.7 shows the average unit prices for the countries studied here. (For Laos no monetary values and for the United States no biophysical data of exports were available and therefore these data are missing in the figure.)

Exports of South American countries mainly consist of raw materials, with low average unit prices of between 150 and 300 USD/t. Chile, with its special situation concerning copper exports, reaches a higher value of 750 USD/t. Average unit prices of imports on the other hand are higher, that is, between 600 and 900 USD/t. Countries in Southeast Asia export goods of much higher economic value per ton. In particular, the exports of the Philippines and Thailand reach average unit prices of 2000 USD/t. Compared with South America, unit prices of imports are lower in Southeast Asia. Average unit prices of exports of the selected OECD countries exceed those of developing countries by far and may be as high as 7500 USD/t for the EU-15 countries and 5000 USD/t for Japan. Thus, the

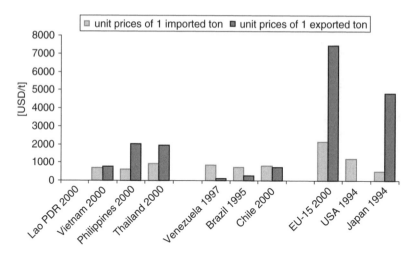

Figure 7.7   Unit prices (USD/t) of imports and exports

EU-15 countries and Japan export final products of high value and can gain higher economic returns with their exports. South American countries have to export disproportionately large amounts of materials and still cannot easily achieve comparable gains from trade.

The figures for the Philippines and Thailand reveal an issue that was not yet acknowledged by biophysical data. Average unit prices in Figure 7.7 suggest that these two countries appear to export a significant share of manufactured goods (see also Schor 2005). In Table 7.1 we compare monetary figures to take a closer look. Venezuela and Chile mainly export raw materials (80 per cent and 90 per cent respectively), whereas the Philippines and Thailand mainly export manufactured or high-technology goods (50 and 65 per cent and 74 and 31 per cent respectively). In the case of Brazil, primary and manufactured exports are more or less balanced. These figures reflect the far greater diversity of the Brazilian economy compared with economies of the other two South American countries, which are more specialized in the extraction and export of one resource.

Trade figures underline the specialization of the two regions in the world economy, which characterizes South American countries as extractive economies that exploit their natural resources for exports and Southeast Asian countries as making use of their abundant labour force to produce labour-intensive manufactured products. High population figures, the significant role of agricultural production and of manufacturing for exports, provide evidence for a pattern of labour-intensive economies.

To derive benefits from production for world markets, it is essential that

Table 7.1   Exports of primary, manufactured and high-technology products

	Primary Exports (% of merchandise exports)	Manufactured Exports (% of merchandise exports)	High-technology Exports (% of manufactured exports)
Philippines	8	50	65
Thailand	22	74	31
Venezuela	89	13	3
Brazil	44	54	19
Chile	80	18	3
USA	14		82
Japan	3		93

Note: No figures were available for Laos or Vietnam.

Source: Data for the year 2002 from UNDP (2004).

production processes and applied technologies exist at a level that does not fall too far below international standards. Integration into the world economy therefore must be accompanied by a transition to industrialized production, at least in those sectors where production is designated for export to world markets. A transition is therefore linked to an increase of overall volumes traded. In this respect, Laos is still at the beginning of its transition, with imports and exports of less than 0.1 t/cap/yr. In Vietnam, traded volumes are below 0.3 t/cap, but considering the relatively higher DE and the higher population in Vietnam, traded volumes in Laos appear even smaller. Traded volumes of all other countries are significantly higher, at 0.6–0.8 t/cap/yr, up to 1.7 t/cap/yr. In relation to the materials used domestically (see Figure 7.8), we see that in Laos and Vietnam less than 3 per cent is imported. All other countries either import around 10 per cent of their materials used or they export a share of 8 per cent or more (with the exception of Chile due to the copper figures already mentioned). Two patterns emerge, again depending on the function in the global division of labour.

7.4   USE OF ENERGY AND LABOUR

The historical transition in Europe from a metabolism merely based on the land and agricultural activities to an industrial metabolism involved a major change in the energy system as well as in materials throughput. With regard to energy, the solar-based energy system was replaced by a fossil-fuel-based energy system (initially dominated by coal and, since the 1950s by oil, gas

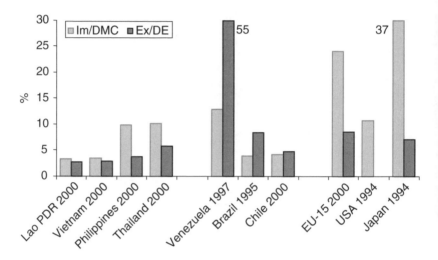

Figure 7.8   Trade intensities: imports/DMC and exports/DE

and electricity). The share of fossil fuels in overall primary energy input can be interpreted as an indicator for the system transition. While in an initial state all energy stems from biomass, during the transitional phase the share of fossil fuels increases rapidly, producing a situation in which energy from biomass amounts to only about 15 per cent to 20 per cent of the energy use (see Chapter 4).

So far, all developing economies in our sample are in the transitional phase with regard to their metabolic profile. Laos has the lowest share of fossil fuel use of 7 per cent (Table 7.2), which supports the pattern of still being a primarily agrarian country. Vietnam and Brazil are still largely dominated by the agricultural context and solar-based energy use with around 70 per cent of the primary energy coming from biomass. Thailand, the Philippines and Chile are in the middle of the energy transition. In these countries, both energy systems exist in parallel and agriculture is to a large extent not affected by the system change. In contrast, fossil fuel use in Venezuela, which has large fossil fuel reserves and fossil fuel production, holds the highest share of 70 per cent and thus is on the brink of establishing a new industrial pattern.

In terms of total energy use, European countries today show a level of energy consumption of about 200 GJ/cap/yr (per capita levels in the United States and Canada are considerably higher). The Latin American economies analysed here are much closer to the European levels of energy use than are the Southeast Asian countries, which have levels of per capita energy use far below those typical for an industrial society and further

Table 7.2   Energy use

	Biomass Use		Fossil Fuel Use		Total (GJ/cap/yr)
	(GJ/cap/yr)	(%)	(GJ/cap/yr)	(%)	
Laos	40	93	3	7	43
Vietnam	23	72	9	28	32
Philippines	26	57	20	43	46
Thailand	31	41	44	59	75
Venezuela	51	32	107	68	158
Brazil	102	73	37	27	139
Chile	50	45	61	55	111

Sources: Krausmann, Erb and Gingrich (2006); fossil fuel use for Laos: UN (2002).

increase has to be expected if these countries would proceed in their transition to an industrial mode of production and consumption.

In the following section, we focus on technical energy used, that is, use of fossil fuels only. We also include human time as a third biophysical dimension. In order to compare Southeast Asia's specialization in labour-intensive production with South America, we focus here on exosomatic energy throughput and human time use for the two regions. The framework used to support this analysis, multi-scale integrated analysis of social metabolism[13] (MSIASM), was introduced by Giampietro and Mayumi (2000a; 2000b) and finally formulated by Giampietro (2003). The approach has been applied to Ecuador (Falconi 2001), Spain (Ramos-Martin 2001), China (Ramos-Martin 2005), and Vietnam (Ramos-Martin and Giampietro 2005). This approach enables us to analyse systemic interrelations between different biophysical variables, that is, energy and time, thus revealing the specific balance achieved by each country in the utilization of limited resources. As a second step, we combine these biophysical and monetary variables.

When considering the social metabolism of economies, it is useful to regard them as complex, adaptive, dissipative systems. This approach implies that an economy is immersed in a continuous process of self-renewal or self-production. In order to better understand this process we use two concepts originating from ecology, namely 'autocatalytic loop' and 'hyper-cycles'. While the first refers to feedback loops within the system that reinforce each other (an output that feeds an input), the latter acknowledges that in a biophysical sense, the whole reproduction of the system depends on its ability to generate the necessary amount of energy within a specialized sub-system (Ulanowicz 1986). In this sense, Ulanowicz, when describing ecosystems as networks of dissipative elements, distinguishes between two main parts: 1) the part responsible for generating the hyper-cycle (that is, activities generating the surplus on which the whole system feeds – in our case, the productive sectors of the economy), and 2) the part representing a pure dissipative structure – in a physical sense, all sectors that do not contribute to the biophysical reproduction of the system but are mere consumptive parts, including the service and government sectors as well as private households.[14] Undertaking an analysis of socioeconomic processes based on energy accounting, we recognize that the maintenance of any social metabolism (in terms of both energy and materials) requires the existence of an autocatalytic loop of useful energy. That is, a certain fraction of energy and of human work must be used to secure the contin-uous flux of energy extracted from the environment and thus ensure that the rest of the system functions. In other words, we characterize this auto-catalytic loop required to maintain the social metabolism of the overall system in terms of a reciprocal relationship between two resources: 1) 'human activity used to control the operation of exosomatic devices' and (2) 'fossil energy used to power exosomatic devices' (Giampietro 1997). We therefore analyse the ability of an economic system to process energy and materials by separating the physically productive sectors (performing the hyper-cycle) from dissipative parts (Ho and Ulanowicz 2005).[15] The amount of energy consumed per unit of time in the productive sector is used as a proxy indicator for the ability of the economic system to provide for the physical basis of all other social or cultural activities. In a metabolic system, the energy flowing into the system is then used to carry out different activities. A fraction of these activities must be directed to guaranteeing the maintenance of this flow, thereby guaranteeing the (re)production of what may later be consumed as final consumption. Of course, this does not mean that activities not directly aimed at the continuation of the energy influx – that is purely dissipative activities – are unimportant. They are indirectly important, for example, insofar as they help to maintain the flow in the long term, for example, through research or education (Giampietro 1997).

Integrated Analysis

In our integrated analysis we use four scale variables, that is, active human time (THA), total energy throughput (TET), human activity in the pro-ductive sectors of the economy (HA_{PW}), and energy throughput in these sectors (ET_{PW}). These variables determine the size or scale of an economy by quantifying the available human time and the amount of technical primary energy used.

In accounting for human time and its allocation to different activities, we start with total human time, that is, the total population multiplied by 8760 hours, representing one year. We then subtract the physiological and social overhead from this 'time budget'. The physiological overhead includes the time spent by each person in self-reproduction, that is, sleeping, eating, personal care, and so on. Social overhead is defined as economically inactive time required for the maintenance of social cohesion, for example, children in childcare and education, the elderly after retirement and leisure time. While the physiological overhead on human time seems to be generally constant in the long term, the social overhead offers more room for decision-taking. The length of time children spend in education and the age at which people enter retirement, or the amount of leisure time made available to individuals are all generally determined by socio-political negotiation. The time remaining after these two subtractions are performed is the amount of time that may potentially be used economically as effective work time. Furthermore, with regard to this fraction of time, a compromise has to be made between activities required to maintain the self-reproduction of the household (for example, cooking, cleaning and childcare) and all economically productive activities. The productive time, that is, work time in agriculture, industry or services, may be spent in subsistence (as is the case in traditional societies in the agricultural sector) or in formal employment. In our analysis, we concentrate on the work time spent in agriculture, industry and services, irrespective of whether the time is spent in subsistence or in market-oriented activities. Total energy throughput refers to the technical primary energy consumption in the economy including fossil fuels and wood fuel.[16] Energy throughput in the sectors producing added-value summarizes primary energy consumed in agriculture (AG), industry (IND), and services and government (SG).

Combining these scale variables results in four intensive variables: exosomatic metabolic rate (EMR) refers to the total primary energy consumed per hour of human time (expressed as joules per hour). While the EMR of the overall system, that is, energy availability per unit of human life time, is considered as an indicator for material standard of living (Giampietro, Bukkens and Pimentel 1997), the sectoral exosomatic metabolic rates in agriculture, industry and services are understood as indicators for capital accumulation in these sectors. A sector showing a higher EMR ratio is, ceteris paribus, endowed with more machinery and tools and is clearly more capital-intensive than a sector showing a lower EMR ratio. What is ignored in this measurement, however, is the efficiency of energy use, where a lower EMR can also indicate a better technology application of the capital invested.[17]

Changes in a scale variable indicate that a system is growing, whereas changes in the intensive variables, when compared over time, are indicative of qualitative rearrangements (that is, development). For instance, a rise in the energy available in relation to human time, that is, a higher EMR, shows increases either in the energy supply to households or the tertiary sector or in the capital accumulation of the economy (that is, the replacement of human work by machine power in the productive sector under consideration).

The biophysical reading, which maps human time against available energy, is complemented by an economic reading, mapping human time against economic value-added. In this economic portrait, the variables for primary energy consumption are replaced by overall GDP and GDP in the productive sectors of the economy under consideration. Resulting intensive variables are economic value-added per unit of active time (GDP/THA) and average return to labour in the productive sectors (GDP_{PW}/THA_{PW}), or economic labour productivity (ELP_i).

Certain case studies (such as Cleveland et al. 1984 and Hall et al. 1986) have found a close correlation between EMR_{PW} and ELP_{PW}.[18] If this hypothesis holds, it follows that economic growth (at least in its earlier stages) implies that the energy consumption in the productive sectors (ET_{PW}) has to grow faster than the time allocated to those sectors (HA_{PW}), resulting in an increase in the energy used per hour of work (EMR_{PW}). This development, at the same time, will increase the ability to invest in capital for future production.

Taken together, the biophysical and economic reading should support a deeper understanding of the functioning of the economies studied, in particular how these economies allocate their resources (time, energy and money) to meet certain ends. We shall investigate whether there is a difference in resource allocation in the two world regions analysed here.

First, we compare an aggregate for Southeast Asia (comprising Thailand, Philippines and Vietnam)[19] with an aggregate for South America (composed of Venezuela, Chile and Brazil). The assessment is performed for two points in time, in the years 1980 and 2000.

Empirical Results

Tables 7.3–7.5 show data for the two aggregated regions using the variables in an integrated analysis as presented before. We distinguish between three levels of analysis: *level n* is the focal level of analysis, in this case the region; *level n-1* shows the disaggregation of human time into working and non-working time, that is, between the hyper-cycle and the dissipative part; *level n-2* explains the composition of working time, by representing the different economic sectors, agriculture (AG), industry (IND), and services and government (SG). We present the extensive variables that define the scale of the system for all three levels: human time (HA_i), the use of technical energy (ET_i), and the economic output (GDP_i) as well as the intensive variables including the requirement of technical energy per hour of working time (EMR$_i$), and the economic productivity of labour (ELP$_i$). We have added the energy efficiency per sector, that is, the relation between the value-added generated and the energy consumption in a sector, measured in USD per gigajoule (GJ).

At *level n*

In terms of population and subsequently available human time the two cases are very similar, starting at a comparable level in 1980 and showing a similar evolution in time until the year 2000. However, in terms of energy consumption, South America is almost double the size of Southeast Asia, but the gap is getting smaller. In fact, total primary energy throughput (TET) has been growing in Southeast Asia at a rate of 4.4 per cent per year, whereas in South America this rate was lower, at 2.5 per cent. As a result, South America shows a level of energy consumption per unit of time that is almost double that of Southeast Asia, yet here too the gap is closing, suggesting that Southeast Asia is catching up in terms of energy intensity (Table 7.3).

At *level n-1*

At this level (Table 7.4), interesting differences show up. For instance, even though the total available time is similar in the two regions, the distribution between working and non-working time is very different. Overall working time (HA$_{PW}$) is much lower in South America than in Southeast Asia but it is growing faster, whereas non-working time is growing at a similar rate. This means that South America is adding new, young, population to the labour force at a faster rate than is the case in Southeast Asia.[20] This is an interesting result that we may use later to explain the particularities of the different sectors.

From the point of view of energy consumption, the two regions show increases in the energy allocated to working activities, but the huge difference between these regions and industrial countries persists in terms of the energy consumption in non-working activities, which is used as a

Table 7.3   Main indicators by region at level n

| | | Southeast Asia | | South America | |
		1980	2000	1980	2000
	Level n				
Total energy throughput	*TET* (PJ/yr)	2629	6466	6581	11158
Total human time	*THT* (Gh/yr)	1288	1897	1295	1838
Exosomatic metabolic rate	*EMR$_{SA}$* (MJ/h)	2.04	3.41	5.08	6.07

Table 7.4   Main indicators by region at level n-1

		Southeast Asia		South America	
		1980	2000	1980	2000
	Level n-1				
Energy throughput PW	ET_{PW} (PJ/yr)	2 060	4 277	4 729	8 546
Energy throughput HH	ET_{HH} (PJ/yr)	569	2 189	1 853	2 612
Human time PW	HA_{PW} (Gh/yr)	142	216	108	183
Human time HH	HA_{HH} (Gh/yr)	1 147	1 681	1 187	1 655
Exosomatic metabolic rate PW	EMR_{PW} (MJ/h)	14.55	19.77	43.83	46.72
Exosomatic metabolic rate HH	EMR_{HH} (MJ/h)	0.50	1.30	1.56	1.58
Gross domestic product	GDP (Million \$/yr)	76 154	185 774	430 799	680 361
Economic labour productivity	ELP (\$/h)	0.54	0.86	3.99	3.72
Economic energy efficiency PW	ELP/EMR_{PW} (\$/GJ)	37.0	43.4	91.1	79.6

Note: GDP in constant 1990 USD (UN Statistics Division 2004).

proxy variable of the material standard of living in the household. While energy-saving technology would probably contribute to a lowering of this variable but still indicate a rise in the standard of living, this is not the case in a developing situation starting from a very low level, where up-to-date technology is often simply not affordable.[21]

It is worth mentioning that Southeast Asia, starting from an EMR_{HH} of 0.5 MJ/h in 1980 (Table 7.4), a third of the level of South America, caught up by the year 2000 and the tendency seems to be stable since that value was growing at an annual rate of 4.7 per cent as compared with 0.1 per cent for South America. A similar pattern is shown for the EMR_{PW}, namely, energy metabolized by the productive sectors, which can be understood as roughly reflecting the level of capital accumulation of these economies and therefore their ability to keep on growing in the future. Since the 1980s, the EMR_{PW} value for Southeast Asia has grown at a yearly rate of 1.5 per cent, whereas in the case of South America this figure was only 0.3 per cent. However, South America still shows more than double the amount of energy metabolized per hour of work. The differences in the 'capitalization' variable are asserted for economic labour productivity, which was only 0.86 USD per hour in Southeast Asia in 2000, but 3.72 USD/hr in South America. Although South America currently operates at a much higher level of economic labour productivity, Southeast Asia is catching up fast because labour productivity has been rising in Southeast Asia but decreasing in South America. This is consistent with the result that in the period analysed, the energy metabolized in working time in South America was growing at a very modest rate. Assuming a close relationship between energy consumption and labour productivity, South America might find itself in a lock-in situation where energy provision in industry and manufacturing cannot keep pace with a rising labour force. It is not likely that gains in energy efficiency will compensate for the decreasing availability of primary energy per working hour.

At *level n-2*

At the level of the economic sectors, the patterns observed above can be explained. A first difference between the two regions is the distribution of working time among sectors. While agriculture employed (formally and informally) 59 per cent of working time in Southeast Asia, it was only 19 per cent in South America in 2000. In both cases, the tendency over time is a decrease of working time devoted to agriculture, more evident, however, in the case of South America. The industry sector (here including also the energy sector and mining sector) accounted for 16 per cent of working time in Southeast Asia and 20 per cent in South America. While in Southeast Asia, this share was growing, in South America it went down from 25 per cent in 1980. The most significant differences occur in the share of the services and government sector. This consumed 31 per cent of the working time in Southeast Asia in the year 2000 and an astonishing 60 per cent in South America. In both cases the share is growing. A plausible explanation for this result is that, Southeast Asia being more rural, it uses the agricultural sector as a cushion for excess labour force, while South America, being more urban, does the same with the services sector (that is, allowing for mainly low value-added services such as informal street selling). In terms of the total amount of working time, we can confirm this hypothesis with data presented in Table 7.5. The regional differences in the agricultural sector or the services sectors persist during the time period studied, and Southeast Asia is catching up with South America only in the case of industry.

In terms of sectoral energy use, only the industry sector is developing along a similar path in the two regions. By contrast, both the agricultural sector, and above all, the services sector is growing much faster in terms of energy use in Southeast Asia than in South America.

Table 7.5 *Main indicators by region at level n-2*

	Southeast Asia		South America	
	1980	2000	1980	2000
Level n-2				
Energy throughput AG	ET_{AG} (PJ/yr)	90.7	149.3	245.0	321.1
Energy throughput IND	ET_{IND} (PJ/yr)	1789.4	3255.3	3540.5	6378.9
Energy throughput SG	ET_{SG} (PJ/yr)	180.1	872.8	943.3	1845.9
Human time AG	HA_{AG} (Gh/yr)	97.8	115.5	29.3	34.9
Human time IND	HA_{IND} (Gh/yr)	16.1	33.8	26.8	37.4
Human time SG	HA_{SG} (Gh/yr)	27.7	67.1	51.7	110.6
Exosomatic metabolic rate AG	EMR_{AG} (MJ/h)	0.93	1.29	8.36	9.20
Exosomatic metabolic rate IND	EMR_{IND} (MJ/h)	110.92	96.31	131.90	170.63
Exosomatic metabolic rate SG	EMR_{SG} (MJ/h)	6.50	13.00	18.23	16.69
Gross domestic product AG	GDP_{AG} (Million \$/yr)	17640	29419	26148	45731
Gross domestic product IND	GDP_{IND} (Million \$/yr)	20053	62603	159966	201083
Gross domestic product SG	GDP_{SG} (Million \$/yr)	32833	85690	215178	391403
Economic labour productivity AG	ELP_{AG} (\$/h)	0.18	0.25	0.89	1.31
Economic labour productivity IND	ELP_{IND} (\$/h)	1.24	1.85	5.96	5.38
Economic labour productivity SG	ELP_{SG} (\$/h)	1.18	1.28	4.16	3.54
Economic energy efficiency AG	ELP/EMR_{AG} (\$/GJ)	194.4	197.01	106.7	142.4
Economic energy efficiency IND	ELP/EMR_{IND} (\$/GJ)	11.2	19.2	45.2	31.5
Economic energy efficiency SG	ELP/EMR_{SG} (\$/GJ)	182.3	98.2	228.1	212.0

Although Southeast Asia is increasingly directing energy into industry and manufacturing, there remains a huge difference in EMR_{IND} at 170 MJ/h in South America and only 96 MJ/h in Southeast Asia in the year 2000. Similar differences apply for the other two sectors. The relevant issue here is that in Southeast Asia, EMR_{IND} decreased at a rate of 0.7 per cent per year. Working time in industry rose faster than energy consumption. The result is very relevant in terms of potential future development. Assuming a strong correlation between EMR and ELP, investment in technology is what drives increases in economic labour productivity (which eventually enhances the overall economic performance of the economy in terms of economic growth). Because such investment relies on present surplus and availability of energy, a decreasing energy/labour ratio should give cause for concern. Should this tendency continue in the medium term, the future development of the region may be dependent upon the inflows of foreign capital, and not be driven by internal dynamics.

Finally, the productivity of labour in the different sectors (which is close to the labour unit costs in any case) helps explain the different roles the two regions play in the world economy. Economic labour productivity (ELP) in the different sectors is much lower in Southeast Asia than in South America. However, in the case of industry, while ELP is decreasing in South America at a rate of 0.5 per cent a year, it is increasing in Southeast Asia at a rate of 1.9 per cent, explained by a transfer of the production of light goods requiring low levels of capital and cheap labour to Southeast Asia.

Looking at the contributions of the individual countries to the aggregates we see that what we define as a Southeast Asian development is to a large degree driven by Thailand, which is characterized by a significantly high increase of EMR_{IND} and EMR_{AS}. In South America, the individual countries more or less follow a similar development, with Chile alone showing a significant increase in EMR_{AS}. Moreover, most countries have a higher energy throughput in the productive sectors. Interestingly, the Philippines show a reverse development, that is, energy throughput in the productive sectors decreases compared with total energy throughput because the increases in energy are consumed by growing population numbers and do not contribute to capital accumulation.

Thailand, once again, underpins the Southeast Asia dynamic with rapid development especially in ELP_{IND}, and GDP per capita. In South America, Chile is revealed to be the success story, with a high increase in GDP per labour time, both in the productive sectors and in the overall economy. In the country aggregate, Chile's improvement is compensated by a reverse development in Venezuela. Venezuela shows a major decrease in overall GDP per human work hour.

Interpretation of the Results: Prospects of Future Development and Possible Constraints

Economic development over time is not only constrained or framed by internal conditions such as resource endowment (natural resources, labour force, knowledge), but is also affected by external constraints upon national economies such as the specific role an economy is playing in the world system, either because of a historical lock-in (that is, enclave and extractive economies), or because national strategies of development have to comply with international economic and political settings (such as World Trade Organization regulations).

In this chapter, we have emphasized the distinctive characteristics of the two trajectories in the transition from an agrarian to an industrialized mode of production adopted by the two regions of Latin America and Southeast Asia during recent decades and have explored how this was reflected in the relevant biophysical variables. With regard to energy availability in the different economic sectors, we see Southeast Asia catching up with a South American region that somehow remains static. However, in both regions, metabolic rates are still very low compared with those of developed economies.

One critical issue in a situation of early industrialization on the path from an agrarian economy concerns how to incorporate new labour force into the economy without lowering the energy available per worker. Usually, this can only be achieved by increasing the total amount of energy consumed by the economy, which in turn will clearly have an impact on the global environment. This result is more clearly in evidence in Southeast Asia, where the labour force is shifting from agriculture (with less consumption of energy per hour) to industry (with higher and increasing consumption of energy per hour of work). In fact, Southeast Asia was unable, despite the huge increase in energy consumption, to achieve the same amount of energy per working hour in industry (EMR_{IND}) in the year 2000 as it did in 1980, simply because labour availability increased at an even higher rate. Due to the high correlation already mentioned between energy and labour productivity, further increases in the economic labour productivity are at risk and the countries may fall into the 'trap' of specialization, meaning they will persist in selling cheap, and ever cheaper labour over time. This is not a strategy that can be maintained in the long run, so we expect to see increases in energy consumption in industry in the future. As we have pointed out above, authors dealing with the deterioration of the terms of trade, such as Sapsford and Balasubramanyam (1994), see diversification as a way out of this trap. But that requires capital accumulation, which, according to our approach of looking at energy metabolism, is reflected in an increase of the amount of energy controlled per hour of work.

On the other hand, the large fractions of population that are still employed in the agricultural sector in the region are likely to be released sooner or later, putting more pressure on these economies to raise the level of energy consumption in order to remain competitive. A similar picture can be seen in South America's low wage part of the service sector, which is used as a buffer when economic performance is not sufficient to absorb additional working population. Here too, the problem is that EMR is much lower than in the industrial sector, as is the productivity of labour, although the large amounts of energy in the industry sector are not coupled with high economic return, reflecting the inefficient use of energy in this sector. This can be seen when looking at how many USD are generated with one GJ of energy spent in that sector. It is astonishing to see that in both cases, agriculture provides four or five times more income per GJ used. For instance, in South America, the agricultural sector in the year 2000 generated 142 USD per GJ consumed, and the services sector 212 USD, whereas the industrial sector generated only 31.5 USD per GJ. These results however, obscure the fact that the energy sector is responsible for making energy available to the rest of the economy, including households, so we cannot interpret this to mean that one option for development is merely to shift economic activities from industry to services. Moreover, the 31.5 USD in the case of industry also reflect the situation of the mining sector, which in the case of Venezuela or Chile accounts for 50 per cent of the GDP of the industrial sector. Zanias (2005) shows that the terms of trade for developing countries deteriorated by 62 per cent between 1920 and 1984. If we take this result into account, it may explain the very low efficiency of energy in generating value-added in these sectors. Also, as we have seen, in these two countries, raw materials exports account for 80 per cent and 90 per cent respectively of total exports. In the case of Thailand or the Philippines, we see a similar high share, but in this case for manufactured goods.

In any case, analysis of both regions suggests that more activity in the industrial sectors will be needed in the future, which will probably imply increases in the energy metabolized, should these two regions prove able to absorb the excess of labour force that is nowadays 'hiding' in the services sector (in South America) and in the agriculture sector (in Southeast Asia).

The cases of South America and Southeast Asia show that both regions today operate at a comparably low level of energy use but will probably increase their metabolic requirements in order to develop their economies and to raise the standard of living of their populations. Such increases in materials and energy use will contribute to further environmental problems such as increasing levels of CO_2 emissions that cause the world climate to deteriorate further. Although it is clear that more efficient technology would promote a more sustainable development in these parts of the world, it is unclear who would pay for such investments. Today, it is increasingly the case that the countries of the semi-periphery begin with activities recently halted in the industrial centres, such as iron and steel or cement production using outdated technologies. It would appear that the objective of growth might override sustainability concerns at least in the short term, because a strategy of growth now and cleaning up later is likely to be pursued.

7.5   TRANSITION IN SOUTHEAST ASIA AND SOUTH AMERICA – TWO TRAJECTORIES?

A regional comparison between South America and Southeast Asia allows us to better understand recent transition processes and related biophysical patterns. We have shown how a transition from an agricultural to an industrial mode of production is reflected in biophysical patterns, but at the same time we see countries in the two regions following different trajectories. We have analysed how these two regions allocate their resources, material, energy and time to certain activities and the extent to which this is determined by the particular role they play in the world economic system.

The three South American countries included in our analysis, Brazil, Chile and Venezuela, are characterized by a high share of urban population, and an agricultural sector of minor importance in terms of employment and contribution to gross domestic product. Although countries such as Brazil are large producers of agricultural goods, production is often mechanized and undertaken on a large-scale basis. South America's role in the world markets is characterized by the provision of raw materials, especially metals and fossil fuels for production in other economies. Extraction of raw materials for exports thus results in higher levels of material extraction compared with Southeast Asian countries and contributes to a negative physical trade balance (that is, these countries are net exporters of their natural resources). This pattern of 'extractive economies' results in a depletion of natural resources, which, rather than being fed into domestic production, are used as exports. Compared with the OECD countries selected in our analysis, use of domestic material resources is still lower because of a considerably lower material standard of living and the marginalized position of sections of the population. Raw materials are exchanged on the world markets for low prices. Thus, specialization on raw material exports usually fails to provide high economic returns and to contribute to employment and household income in a significant way. The specific role of South

American countries in the world economy is rooted in a high degree of specialization, primarily related to one main economic activity, and therefore a high dependency on world markets and international prices. The only exception among the countries we studied in detail is Brazil, which shows a greater diversification of economic activities. In terms of labour and energy, South American countries use high amounts of energy particularly in the productive sectors, and comparatively little human labour time, again reflecting the specialization in export-oriented primary activities that are more capital- than labour-intensive. The South American countries have achieved a certain level of wealth above that of the Southeast Asian economies. However, this higher economic income is less evenly distributed among the population. South America is without doubt already in a process of transition with a number of countries exhibiting industrialized patterns, such as a high share of fossil fuel use compared with biomass use and already having a high degree of integration into world markets through their very specific functions within the world economy.

Southeast Asia, on the other hand, is different. In the countries we included in our assessment, namely Thailand, Laos, the Philippines and Vietnam, the agricultural sector is still important for economic growth but even more so for employment. A considerable part of agricultural production takes place in subsistence forms outside the market. This results in a lowering of the average national income but is otherwise effective in providing food security. Most of the agricultural holdings are small scale and family farmers produce on small plots of sometimes less than a hectare. Industrialization of the agricultural sector is a marginal phenomenon and is restricted to specific cash crop production on plantations.

Domestic material extraction is much lower than that of South American countries and even lower again in relation to the selected OECD countries, and in terms of their physical trade, these countries are net-importers of natural resources. Economic activities in the Southeast Asian economies show a higher diversity than those in South America. To take Thailand as an example, the country still has a large agricultural sector with a considerable share of subsistence production especially in Northwest Thailand. At the same time, industrial production of finished goods such as textiles or consumer electronics and activities in the tertiary sector have found their place in the economic structure. Given the very high population densities and the relative abundance of labour, Southeast Asia as a whole follows a different trajectory in the process of transition compared with South America. While the exploitation of natural resources by Third Countries is not a dominant phenomenon, it is the cheap labour that determines Southeast Asia's position in the world economy. This was confirmed by taking a closer look at the energy/labour relations within Southeast

Asian economies, which show an extensive use of labour time, especially in the productive sectors, but significantly less energy throughput. Since value-added per labour hour is much lower than in South America, we can conclude that Southeast Asian countries are exploiting their labour resources rather than their material resources. Although by development standards and in terms of per capita GDP, Southeast Asia clearly lags behind South America, GDP growth rates since the 1980s have been considerably higher in Southeast Asia, which indicates that the economic diversification strategy is probably more successful than the sole orientation on raw material exports.

In Southeast Asia, we may anticipate very rapid economic development due to the growth potential of the large agricultural population, partly in terms of consumers ready to supply the industrial development with new labour force and new consumers. This development is already visible in Thailand and to a lesser degree in Vietnam. Changes in the agricultural sector were a decisive factor in the historical transformation of the industrialized countries of today. The future pathway taken by the agricultural population in Southeast Asia will also help determine whether the region will experience development of a sustainable kind.

Laos, uniquely among the countries studied, is still right at the beginning of a transition process, with patterns closely resembling an agricultural economy, that is, a biomass-based economy and marginal integration into world markets. Laos has a very low population density (much closer to South American levels) and has experienced almost no industrialization. The economy is characterized by a high share of subsistence production but hardly any significant manufacturing or industrial production. However, Laos is rich in natural resources, including hydro power, timber and minerals, and also has a wealth of biodiversity and biological resources. In a move recalling the historical development of many South American countries, Laos has started to exploit its natural resources with the help of foreign investment. Whether Laos will follow the development trajectory of Southeast Asian countries or instead pursue a South American pattern, complementing its Southeast Asian neighbour states, remains to be seen.

In terms of future development and sustainable development in particular, the two regions are confronted with different opportunities and constraints. South American countries are highly specialized in raw material extraction and export, with a much smaller agricultural sector but a high share of urban population. Future economic development in this region must be based in the primary sector where it is difficult to increase economic gains because the relative prices for raw materials are constantly declining. At the same time, there is intense competition from other countries supplying the same products, and technology-related gains are not expected to compensate for environmental problems related to resource extraction and decreasing natural stocks and deposits. South American countries might be able to introduce diversification of their economic activities and thus build up other sectors that reduce the dependency on raw material extraction and exports. This would also help relieve environmental pressures directly related to these primary economic activities, such as land-use change, water pollution and waste generation. Brazil has the potential to serve as a role model for such a development.

Southeast Asia, on the other hand, is currently building up an industrial and manufacturing sector based on a high input of labour time while still retaining a significant agricultural sector, which provides a buffer to absorb workforce affected by times of crisis. Southeast Asian economies are therefore engaged in a development process that is reducing risks and decreasing dependency upon world markets considerably, especially with regard to food availability and food security. Despite this, some of the countries of Southeast Asia are still among the poorest countries in the world, with a very low material standard of living. Increasing the affluence of people and finding solutions to some of the pressing problems facing the region will depend upon whether the countries of the region manage to use the abundant labour force in a sustainable way. In terms of resource use, Southeast Asian countries may not be exploiting their material resources but they will probably be confronted in the future with problems related to the intensive use of their land resources. For data tables for the regions discussed in this study, see Appendix 7.1 at the end of this chapter.

ACKNOWLEDGEMENTS

The comparative analysis presented in this chapter was made possible by the research project 'GEWIN – Social Knowledge and Global Sustainability' funded by the Austrian Ministry of Science Programme for Cultural Landscape Research (KLF). The country studies data analysed in our chapter were conducted in two EU-funded projects: 'Amazonia 21 – Operational Features for Managing Sustainable Development in Amazonia', an FP4 INCO project, 1998–2001 and 'SEAtrans: Southeast Asia in Transition. Social Transitions, Environmental Impacts and Policies for Sustainable Development', an FP5 INCO project, 2000–03.

The authors wish to thank Willi Haas, Clemens Grünbühel, Julia Lutz, Niels Schulz and Harald Wilfing for many discussions within the GEWIN project as well as other members of the Institute of Social Ecology for fruitful discussions. We are especially grateful to Marina Fischer-Kowalski and

Helmut Haberl for detailed comments on earlier versions of this chapter and their efforts in enabling the production of this book.

NOTES

1. The staple theory of growth was developed in the 1920s by Harold Innis to explain the rapid economic growth of Canada. The main idea in this export-led growth model is that countries with abundant resources may specialize in exporting staple goods, which will have a positive effect on real growth rates for GDP per capita (Altman 2003). North (1955) even suggested the use of this theory to explain how non-agricultural or raw-material-based exports could positively affect economic growth.
2. The Heckscher-Ohlin trade model (Heckscher 1919), states that comparative advantage in production and export is driven by the endowment with production factors (natural resources, labour, technology or knowledge). Therefore, countries specialize in production processes that make intensive use of the abundant factor and will import products, based on intensive use of resources that are comparably scarce in their own economies.
3. The database Bloch and Sapsford (1997) used, as most of the studies on the topic, makes use of the primary price series, an index of the prices of 24 internationally traded non-fuel primary commodities as calculated by the World Bank and later updated by the IMF.
4. Laos and Vietnam show comparably high shares (76 per cent and 67 per cent respectively) of labour force engaged in agricultural activities. In Thailand, 56 per cent of the labour force are engaged in agricultural activities and 40 per cent in the Philippines. The contribution to GDP is between 9 per cent in Thailand and 52 per cent in Laos.
5. Venezuela has only 8 per cent of its labour force occupied in agriculture while in Brazil and Chile around 15 per cent of the labour force is agricultural. Contribution to GDP is between 5 per cent in Venezuela and 7 per cent in Chile.
6. At constant 1990 prices in USD.
7. For example, maize (SA: 3.5 t/ha/yr, SEA: 2.5 t/ha/yr), sugar cane (SA: 65 t/ha/yr, SEA: 50–60 t/ha/yr), potatoes (SA: 17 t/ha/yr, SEA: 12 t/ha/yr), and even paddy rice (SA: 3–5 t/ha/yr, SEA: 2.5–4 t/ha/yr).
8. Domestic extraction (DE) = all materials (solid, liquid and gaseous excluding water and air) extracted from the national environment for further use in production or consumption processes (Eurostat 2001, pp. 17 and 21). Domestic material consumption (DMC) = DE + imports – exports = total amount of material directly used in an economy (Eurostat 2001, p. 36). DMC can be understood as the 'domestic waste potential' that is, all materials used and consumed both in production and consumption processes (Weisz, Krausmann and Amann et al. 2006). Physical trade balance (PTB) = imports – exports (Eurostat 2001, p. 36).
9. Material flow accounts for the three South American and four Southeast Asian economies have been established within two EU-funded INCO-DEV research projects called 'Amazonia 21' and 'Southeast Asia in Transition'. The MFA for Chile was conducted by Giljum (2004).
10. Vietnam is one of the world's leading producers of anthracite coal. In 2004, Vietnam ranked sixth in production of crude petroleum in the Asia and Pacific region (*Oil & Gas Journal* 2004). Thailand has small resources of coal, natural gas and crude petroleum.
11. http://www.aseansec.org/10367.htm accessed 21 March 2006.
12. Domestic extraction in China, India and Indonesia was as low as 3–3.2 t/cap/yr in the year 2000. Even with considerable amounts of imports (and fewer exports), the domestic consumption will be far below 8 t/cap/yr.
13. In Giampietro's work (2003) he used the term 'societal metabolism'. But for reasons of coherence throughout the book we use 'social metabolism' instead.
14. Although we characterize households here as mainly consumptive units, this might be different in the context of subsistence economies. Also, such a characterization ignores

the fact that households have certain functions for their members including biological reproduction, childcare and other social services.

15. While the hyper-cycle directly maintains the basic requirement of metabolism, the dissipative parts have their main function in systems organization and knowledge acquisition in order to allow for change.

16. In contrast to other chapters in this book, we restrict our energy analysis to the exosomatic part of energy use, separating the part of the overall energy flow required to nourish humans (endosomatic flows) from all other forms of energy. We understand the hyper-cycle as the part related to the capacity of the economy to grow and to support a better material standard of living of the population, while the nutritional energy flow would be related to demographic development, the health and nutritional condition of the labour force and their ability to work. Therefore, the energy figures used here are somewhat lower than comparable figures in other chapters that include nutritional energy from biomass.

17. The influence of technology and efficiency could be addressed at a later stage in the energy analysis, when the primary energy input is distributed to final demand categories and finally to useful energy (Hall, Cleveland and Kaufmann 1986). This has, however, not been the objective of this chapter.

18. The correlation between energy availability in economic sectors and labour productivity in these sectors found by Cleveland et al. in their analysis for the US economy was also found for Spain (Ramos-Martin 2001) and for Ecuador (Falconi 2001).

19. Laos is not included in the Southeast Asian aggregate as its socioeconomic structure and development is one of a rather agrarian economy before or right at the edge of a transition to an industrial mode of production.

20. This difference is mainly due to past population growth and age structure.

21. Even in industrialized countries, we have to acknowledge that efficiency gains and lower prices might cause a rise in consumption contributing to overall rise in resource use, the well-documented rebound effect (F. Jevons 1990; W.S. Jevons 1865; Giampietro 1994; Herring 1999).

REFERENCES

Adriaanse, A., Stefan Bringezu, A. Hammond, Yuichi Moriguchi, Eric Rodenburg, Don Rogich and Helmut Schütz (1997), *Resource Flows: The Material Basis of Industrial Economies*, Washington DC: World Resources Institute.

Altman, Michael (2003), 'Staple Theory and Export-led Growth: Constructing Differential Growth', *Australian Economic History Review*, **43**(3), 230–55.

Amin, Samir (1976), *Unequal Development*, New York: Monthly Review Press.

Bairoch, Paul (1975), *The Economic Development of the Third World Since 1900*, London: Methuen.

Baran, Paul (1957), *The Political Economy of Growth*, New York: Monthly Review Press.

Bloch, H. and David Sapsford (1997), 'Some Estimations of Prebisch and Singer Effects on the Terms of Trade between Primary Producers and Manufacturers', *World Development*, **25**(11), 1873–84.

Bruton, Henry J. (1998), 'A Reconsideration of Import Substitution', *Journal of Economic Literature*, **36**(2), 903–36.

Bunker, Stephen G. (1985), *Underdeveloping the Amazon: Extraction, Unequal Exchange, and the Failure of the Modern State*, Chicago: Chicago University Press.

Cardoso, Fernando H. and Enzo Faletto (1979), *Dependency and Development in Latin America*, Los Angeles: University of California Press.

Cleveland, Cutler J., Robert Costanza, Charles A.S. Hall, Robert K. Kaufmann and David I. Stern (1984), 'Energy and the U.S. Economy: A Biophysical Perspective', *Science*, **225**, 890–97.

Crowson, Philip (2001), *Minerals Handbook 2000–2001. Statistics and Analyses of the World's Minerals Industry*, Kent: Mining Journal Books Ltd.

Dosal, Paul J. (1993), *Doing Business with the Dictators: A Political History of United Fruit in Guatemala, 1899–1944*, Wilmington, DE: Scholarly Resources Books.

Ekins, Paul, Carl Folke and Robert Costanza (1994), 'Trade, Environment and Development: The Issue in Perspective', *Ecological Economics*, **9**(1), 1–98.

Emmanuel, A. (1972), *Unequal Exchange: A Study of the Imperialism of Trade*, New York: Monthly Review Press.

Eurostat (2001), *Economy-wide Material Flow Accounts and Derived Indicators. A Methodological Guide*, Luxembourg: Eurostat, European Commission, Office for Official Publications of the European Communities.

Falconi, Fander (2001), 'Integrated Assessment of the Recent Economic History of Ecuador', *Population and Environment*, **22**(3), 257–80.

FAO (2004), *FAOSTAT 2004, FAO Statistical Databases: Agriculture, Fisheries, Forestry, Nutrition*, Rome: FAO.

Fischer-Kowalski, Marina and Christof Amann (2001), 'Beyond IPAT and Kuznets Curves: Globalization as a Vital Factor in Analysing the Environmental Impact of Socio-Economic Metabolism', *Population and Environment*, **23**(1), 7–47.

Frank, André G. (1967), *Capitalism and Underdevelopment in Latin America*, New York: Monthly Review Press.

Giampietro, Mario (1994), 'Using Hierarchy Theory to Explore the Concept of Sustainable Development', *Futures*, **26**, 616–25.

Giampietro, Mario (1997), 'Linking Technology, Natural Resources, and the Socioeconomic Structure of Human Society: A Theoretical Model', *Advances in Human Ecology*, **6**, 75–130.

Giampietro, Mario (2003), *Multi-scale Integrated Analysis of Agroecosystems*, Boca Raton, London: CRC.

Giampietro, Mario and Kozo Mayumi (2000a), 'Multiple-scale Integrated Assessment of Societal Metabolism: Integrating Biophysical and Economic Representation Across Scales', *Population and Environment*, **22**(2), 155–210.

Giampietro, Mario and Kozo Mayumi (2000b), 'Multiple-scale Integrated Assessment of Societal Metabolism: Introducing the Approach', *Population and Environment*, **22**(2), 109–54.

Giampietro, Mario, Sandra G.F. Bukkens and David Pimentel (1997), 'Linking Technology, Natural Resources, and the Socioeconomic Structure of Human Society: Examples and Applications', *Advances in Human Ecology*, **6**, 131–200.

Giljum, Stefan (2004), 'Trade, Material Flows and Economic Development in the South: The Example of Chile', *Journal of Industrial Ecology*, **8**(1–2), 241–61.

Hall, Charles A.S., Cutler J. Cleveland and Robert K. Kaufmann (1986), *Energy and Resource Quality, The Ecology of the Economic Process*, New York: Wiley Interscience.

Heckscher, Eli (1919), 'The Effect of Foreign Trade on the Distribution of Income', *Ekonomisk Tidskrift*, 497–512.

Herring, Horace (1999), 'Does Energy Efficiency Save Energy? The Debate and its Consequences', *Applied Energy*, **63**, 209–26.

Ho, M.W. and Robert Ulanowicz (2005), 'Sustainable Systems as Organisms?', *Biosystems*, **82**(1), 39–51.

Innis, Harold A. (1930), *The Fur Trade in Canada: An Introduction to Canadian Economic History*, New Haven, CT: Yale University Press.

Jevons, F. (1990), 'Greenhouse: A paradox', *Search*, **21**(5).

Jevons, W.S. (1865), *The Coal Question – An Inquiry Concerning the Progress of the Nation, and the Probable Exhaustion of our Coal-mines*, Houndsmills, Basingstoke, Hampshire: Palgrave.

Krausmann, Fridolin, Karl-Heinz Erb and Simone Gingrich (2006), Personal communication.

Lenin, Vladimir I. [1917] (1971), *Imperialism, The Highest State of Capitalism*, New York: International Publishers.

Luxemburg, Rosa [1913] (2003), *The Accumulation of Capital: A Contribution to an Economic Explanation of Imperialism*, London, New York: Routledge.

Mandel, Ernest (1968), *Marxist Economic Theory*, New York and London: Monthly Review Press.

Matthews, Emily, Christof Amann, Marina Fischer-Kowalski, Stefan Bringezu, Walter Hüttler, René Kleijn, Yuichi Moriguchi, Christian Ottke, Eric Rodenburg, Don Rogich, Heinz Schandl, Helmut Schütz, Ester van der Voet and Helga Weisz (2000), *The Weight of Nations: Material Outflows from Industrial Economies*, Washington, DC: World Resources Institute.

Muradian, Roldan and Joan Martinez-Alier (2001), 'Trade and the Environment: From a "Southern" Perspective', *Ecological Economics*, **36**(2), 281–97.

North, Douglass C. (1955), 'Location Theory and Regional Economic Growth', *Journal of Political Economy*, **63**, 243–58.

Ohlin, Bertil (1933), *Interregional and International Trade*, Cambridge, MA: Harvard University Press.

Oil & Gas Journal (2004), 'Worldwide Look at Reserves and Production', *Oil & Gas Journal*, **102**(47), 22–3.

Overton, Mark and Bruce M.S. Campbell (1999), 'Statistics in Production and Productivity in English Agriculture, 1086–1871', in Bas J.P. van Bavel and Erik Thoen (eds), *Land Productivity and Agro-systems in the North Sea Area (Middle Ages–20th Century). Elements for Comparison*, Brussels: Brepols, pp. 189–209.

Ozawa, T. (2006), 'Foreign Direct Investment and Economic Development', *Transnational Corporations*, **1**(1), 27–54.

Porter, Michael (1990), *The Competitive Advantage of Nations*, New York: Free Press.

Prebish, Raul (1950), *The Economic Development of Latin America and its Principal Problems*, New York: United Nations Economic Commission for Latin America.

Prebish, Raul (1959), 'Commercial Policy in Underdeveloped Countries', *American Economic Review*, **49**, 251–73.

Ramos-Martin, Jesus (2001), 'Historical Analysis of Energy Intensity of Spain: From a "Conventional View" to an "Integrated Assessment"', *Population and Environment*, **22**(3), 281–313.

Ramos-Martin, Jesus (2005), 'Complex Systems and Exosomatic Energy Metabolism of Human Societies', PhD thesis, Autonomous University of Barcelona.

Ramos-Martin, Jesus and Mario Giampietro (2005), 'Multi-scale Integrated Analysis of Societal Metabolism: Learning from Trajectories of Development and Building Robust Scenarios', *International Journal of Global Environmental Issues*, **5**(3/4), 225–63.

Ricardo, D. (1817), *On the Principles of Political Economy and Taxation*, London: John Murray.

Ritchie, Bryan K. (2002), *Foreign Direct Investment and Intellectual Capital Formation in Southeast Asia*, Paris: OECD Development Centre.

Rock, Michael T., David P. Angel and Tubagus Feridhanusetyawan (1999), 'Industrial Ecology and Clean Development in East Asia', *Journal of Industrial Ecology*, 3(4), 29–42.

Sapsford, David and V.N. Balasubramanyam (1994), 'The Long-run Behavior of the Relative Price of Primary Commodities: Statistical Evidence and Policy Implications', *World Development*, 22(11), 1737–45.

Schandl, Heinz and Nina Eisenmenger (2006), 'Regional Patterns in Global Resource Extraction', *Journal of Industrial Ecology*, 10(4).

Schor, Juliet B. (2005), 'Prices and Quantities: Unsustainable Consumption and the Global Economy', *Ecological Economics*, 55(3), 309–20.

Shannon, Thomas R. (1996), *An Introduction to the World-System Perspective*, Colorado, USA; Oxford, UK: Westview Press.

Singer, H.W. (1950), 'The Distribution of Gains between Investing and Borrowing Countries', *American Economic Review*, 40(2), 473–85.

Smith, Adam (1776), *An Inquiry into the Nature and Causes of the Wealth of Nations*, Dublin: Whitestone.

Stiglitz, Josef E. (2002), *Globalization and its Discontents*, New York: W.W. Norton & Co.

Ulanowicz, Robert (1986), *Growth and Development: Ecosystem Phenomenology*, New York: Springer.

UN (2002), *Energy Statistics Yearbook 1999*, New York: United Nations, Department of Economic and Social Affairs.

UNCTAD (2005), *Foreign Direct Investment Database*, http://stats.unctad.org/fdi/.

UNDP (2004), *Human Development Report 2004, Cultural Liberty in Today's Diverse World*, New York, NY: United Nations Development Programme.

United States Bureau of Mines (1987), *An Appraisal of Minerals Availability for 34 Commodities. Bulletin 692*, Washington, DC: US Government Printing Office.

UN Statistics Division (2004), *National Accounts Main Aggregates Database*, http://unstats.un.org/unsd/snaama/Introduction.asp.

Wallerstein, Immanuel (1979), *The Capitalist World Economy*, Cambridge: Cambridge University Press.

Weisz, Helga, Fridolin Krausmann and Sirichet Sangarkan (2006), *Resource Use in a Transition Economy: Material and Energy Flow Analysis for Thailand*, Wieu, Austria: unpublished manuscript.

Weisz, Helga, Fridolin Krausmann, Christof Amann, Nina Eisenmenger, Karl-Heinz Erb, Klaus Hubacek and Marina Fischer-Kowalski (2006), 'The Physical Economy of the European Union: Cross-country Comparison and Determinants of Material Consumption', *Ecological Economics*, 58(4), 676–98.

Weisz, Helga, Fridolin Krausmann, Nina Eisenmenger, Christof Amann and Klaus Hubacek (2005), *Development of Material Use in the European Union 1970–2001. Material Composition, Cross-country Comparison and Material Flow Indicators*, Luxembourg: Eurostat, Office for Official Publications of the European Communities.

Zanias, George P. (2005), 'Testing for Trends in the Terms of Trade Between Primary Commodities and Manufactured Goods', *Journal of Development Economics*, 78(1), 49–59.

APPENDIX 7.1 DATA TABLES

Year 2000	Land Area (1000 km²) (FAO 2004)	Population (1000 capita) (FAO 2004)	Population Density (capita/km²) (own calculation)	GDP (million USD)* (UN 2004)	GDP/capita USD*/capita (own calculation)	HDI (2001) (UNDP 2004)	Gini Index (1990–2000) (UNDP 2004)
World	130.667	6 070.586	46				
Southeast Asia	4.360	522.121	120	562.187	1.077	0.722[e]	0.406[f]
Lao PDR	231	5.279	23	1.415	268	0.525	0.370
Vietnam	325	78.137	240	13.433	172	0.688	0.361
Philippines	298	75.653	254	59.822	791	0.751	0.462
Thailand	511	62.806	123	132.042	2.102	0.768	0.414
South America	17.276	345.155	20	1 050.730	3.044	0.777[d]	0.506[f]
Venezuela	882	24.170	27	59.355	2.456	0.775	0.495
Brazil	8.459	170.406	20	570.937	3.350	0.777	0.607
Chile	749	15.211	20	62.101	4.083	0.831	0.567
EU-15	3.131	376.722	120	8 632.720	22.915	0.924[f]	0.299[f]
USA	9.159	283.230	31	7 968.520	28.134	0.937	0.408
Japan	365	127.096	349	3 492.799	27.482	0.932	0.249

Year 2000	GDP Per Sector			Labour Force Per Sector		Foreign Direct Investment	
	Agriculture[a] (%) UN 2004	Industry and Manufacturing[b] (%) UN 2004	Services[c] (%) UN 2004	Agriculture (%) FAO 2004	Industries and Services (%) FAO 2004	FDI Inflows (million USD) UNCTAD 2005	FDI Outflows (million USD) UNCTAD 2005
World							
Southeast Asia	13	40	47	53	47		
Lao PDR	52	23	26	76	24	0.34	10
Vietnam	29	33	38	67	33	1.289	–
Philippines	19	35	46	40	60	1.345	–108
Thailand	9	36	55	56	44	3.350	–22
South America	8	34	58	18	82		
Venezuela	5	53	42	8	92	4.701	0.521
Brazil	7	33	60	17	83	32.779	2.282
Chile	6	36	59	16	84	4.860	3.987
EU-15	3	27	70	4	96		
USA	1	23	75	2	98		
Japan	2	36	63	4	96		

Notes: * At constant 1990 prices in US Dollars. USD are converted to constant prices by using the annual period-average exchange rate of the base year.
a Agriculture, hunting, forestry, fishing.
b Mining, manufacturing, utilities, construction.
c Wholesale, retail trade, restaurants and hotels, transport, storage and communication, other activities, unspecified.
d Latin America and the Caribbean.
e East Asia and the Pacific.
f Own calculations–mean value of countries in the region.

Material flow data		Lao PDR 2000 (Project SEAtrans)	Vietnam 2000 (Project SEAtrans)	Philippines 2000 (Project SEAtrans)	Thailand 2000 (Project SEAtrans)	Venezuela 1997 (Project Amazonia21)	Brazil 1995 (Project Amazonia22)	Chile 2000 (Giljum 2004)	EU-15 2000 (Weisz et al. 2006)	USA 1994 (Adriaanse et al. 1997; Matthews et al. 2000)	Japan 1994 (Adriaanse et al. 1997; Matthews et al. 2000)
DE total	t/cap/yr	3.5	9.7	7.3	11.4	14.8	14.9	41.4	13.0	21.0	10.2
DE biomass	t/cap/yr	2.3	1.4	1.7	2.9	4.0	8.9	3.3	3.8	3.9	0.9
DE fossil fuels	t/cap/yr	0.0	0.4	0.0	0.7	8.6	0.3	0.2	1.9	6.4	0.1
DE minerals	t/cap/yr	1.1	8.0	5.6	6.8	2.2	5.7	37.9	7.3	10.8	9.2
Imports total	t/cap/yr	0.1	0.3	0.8	1.2	1.0	0.6	1.7	3.8	2.5	5.5
Exports total	t/cap/yr	0.1	0.3	0.3	0.7	8.2	1.3	2.0	1.1	0.0	0.7
PTB total	t/cap/yr	0.0	0.1	0.5	0.5	-7.2	-0.7	-0.2	2.6	2.5	4.8
PTB biomass	t/cap/yr	-0.1	0.0	0.1	-0.2	0.2	0.0	-0.4	0.2	0.0	0.7
PTB fossil fuels	t/cap/yr	0.0	-0.2	0.3	0.6	-6.5	0.2	1.1	1.8	1.9	3.2
PTB minerals	t/cap/yr	0.0	0.0	0.0	0.1	-0.6	-0.9	-0.9	0.6	0.2	1.2
PTB products	t/cap/yr	0.0	0.2	0.1	0.0	-0.3	-0.1	0.0	0.0	0.4	0.4
DMC total	t/cap/yr	3.5	9.8	7.8	11.9	7.6	14.2	41.2	15.6	23.6	15.0
DMC biomass	t/cap/yr	2.3	1.4	1.8	2.6	4.2	9.0	2.8	4.0	3.9	1.6
DMC fossil fuels	t/cap/yr	0.1	0.2	0.3	1.3	2.1	0.6	1.3	3.8	8.3	3.3
DMC minerals	t/cap/yr	1.1	8.0	5.6	7.9	1.6	4.8	37.0	7.9	11.0	10.4
DMC products	t/cap/yr	0.0	0.2	0.1	0.0	-0.3	-0.1	0.0	0.0	0.4	0.4
Unit price of one imported ton	USD/ ton	n/a	683	636	931	864	755	831	2 132	1 203	488
Unit price of one exported ton	USD/ ton	n/a	787	2 006	1 940	136	272	751	7 448	n/a	4 809
Im/DMC	%	3%	3%	10%	10%	13%	4%	4%	24%	11%	37%
Ex/DE	%	3%	3%	4%	6%	55%	8%	5%	9%	0%	7%

Energy use		Lao PDR 2000 (Project SEAtrans)	Vietnam 2000 (Project SEAtrans)	Philippines 2000 (Project SEAtrans)	Thailand 2000 (Project SEAtrans)	Venezuela 1997 (Project Amazonia21)	Brazil 1995 (Project Amazonia22)	Chile 2000 (Giljum 2004)	EU-15 2000 (Weisz et al. 2006)	USA 1994 (Adriaanse et al. 1997; Matthews et al. 2000)	Japan 1994 (Adriaanse et al. 1997; Matthews et al. 2000)
Biomass use	GJ/cap/yr	40	23	26	31	51	102	50	n/a	n/a	n/a
Biomass share	%	93%	72%	57%	41%	32%	73%	45%	n/a	n/a	n/a
Fossil fuel use	GJ/cap/yr	3	9	20	44	107	37	61	n/a	n/a	n/a
Fossil fuel share	%	7%	28%	43%	59%	68%	27%	55%	n/a	n/a	n/a
Total energy use	GJ/cap/yr	43	32	46	75	158	139	111	n/a	n/a	n/a

8. Conclusions: likely and unlikely pasts, possible and impossible futures

Marina Fischer-Kowalski, Helmut Haberl and Fridolin Krausmann

8.1 WHAT DO WE NEED TO LEARN?

This book aims to present an innovative way of looking on humanity's development in the last few centuries by focusing on its biophysical aspects. It complements more traditional views on the recent past, such as those developed within history, sociology, economics, ecology, meteorology, or geophysics, with an interdisciplinary approach that is driven by the quest to understand the interrelations between these domains. This is done by looking at changes in the interrelations between societies and their natural environment during transitions from the agrarian to the industrial socioecological regime. In particular, we analyse changes in patterns in socioeconomic metabolism and land use as key elements of socioecological systems.

Our core message is the following: on the global level, we are still in the midst of a transition that started more than 300 years ago, a transition from an agrarian – that is, land-based – to an industrial socioecological regime based on fossil fuels. Regions in the global North and regions in the global South are at different points of this ongoing transition. Global interdependencies modify the process so that it is not simply repetitive across time and regions. When we strive for a further transition, towards a more sustainable society, as a transformation of our current patterns of living, of production and consumption, we need to be aware that roughly two-thirds of the people on Earth living on more than two-thirds of the world's land area are still pretty close to the traditional agrarian regime, although these parts of the world are mostly rapidly being transformed towards the industrial mode. This industrial mode may promise more concerning quality of life, but it is certainly not more sustainable. Even worse, those promises are unlikely to be kept.

Somehow, while the industrial socioecological regime dominates the world, all countries far enough advanced in this direction resemble, functionally speaking, the urban centres of the preceding agrarian regime. They increasingly function as centres of knowledge management and consumption, drawing upon production activities in the global peripheries. What was once only true for agricultural produce now increasingly also holds true for industrial products: they are manufactured in the peripheries, where much of the raw materials and energy are being extracted. The former function of agrarian cities as a population sink has also been taken over by industrial countries: the OECD core, and Europe in particular, requires a steady inflow of population from the peripheries to maintain its population size and density. Even intellectually, there is an important commonality: like urban populations in the past, the inhabitants of the advanced countries are almost completely unaware of the world outside that supports them.

This gives rise to similar interpretative mistakes: as the citizens of Ancient Rome or 18th-century London may have tended to believe that the age of agriculture was at an end, having given way to an age of cities, so contemporary social science speaks about 'post-industrial society' (Bell 1973), 'affluent society' (Galbraith 1958), the 'end of labour society' (Offe 1984) or 'information society' (Webster 1995). Such notions overlook the deep functional interdependencies between what can be observed within one's immediate horizon – to which such descriptions may apply – and the rest of the world that looks very different, housing many of those processes that have disappeared in the centres.

Looking at societies from a biophysical perspective renders such misperceptions less likely. In economic terms – that is, in socioeconomic descriptions that use money as their common denominator – all processes happening in the core are greatly magnified. In 2003, 81 per cent of global GDP was generated in the OECD countries where only 18 per cent of the world population lived (FAO 2004; UN Statistics Division 2004). The weighing mechanism of money is by far the most skewed: if importance in the world is attributed according to money, everything that happens outside the core and its institutions is quite unimportant. Biophysical measures provide different distributions of importance: the primary energy needed to sustain an average citizen in one of the 50 poorest countries of the world (together containing 11 per cent of the global population) is by a factor of 7 smaller than the energy required for an average inhabitant of an OECD country, available cropland area by a factor of 2 and the food required only by a factor of 1.7. Average human life expectancies vary between countries by no more than a factor of 2.5 (UNDP 2004). These are the basic characteristics and necessities of human life, and these are the resources that are limited on our planet Earth. When we look at societies and social life in these categories, we are able not only to make use in our analysis of some natural science methods but also to gain an impression of some basic equalities between humans across time and space.

In contrast to this, monetary measures such as per capita GDP differ from average citizens in poor to average citizens in rich countries by a factor of 50 (Maddison 2001). Money, even if corrected for inflation and purchasing power to become 'real values', is inherently unlimited. This is what makes it such a convenient medium and allows for so much dynamics in social life. But if monetary values illuminate your worldview, the rich areas are well lit, and the areas with little money are dark, indeed almost invisible.

8.2 LESSONS FROM EUROPE'S PAST: THE METABOLIC PROFILE OF THE AGRARIAN SOCIOECOLOGICAL REGIME

Primary Energy Use

An agrarian socioecological regime is what Sieferle (1997) terms a 'controlled solar energy system': practically all primary energy input originates from land use, that is, from using biologically productive land as cropland, grazing area or for forestry. The deliberate transformation of terrestrial ecosystems into managed landscapes is the central human activity with respect to energy provision. Empirical evidence suggests that in fully developed European agrarian regimes (such as the United Kingdom or Austria in the mid- and late 18th century respectively), primary energy input on a national level typically ranged between 50 and 80 gigajoules per capita per year (GJ/cap/yr).[1] Practically all this energy came in the form of biomass, whereas wind and water power played only a minor role quantitatively (less than 1 per cent in each of the case studies analysed so far).[2] Some 3–10 GJ/cap/yr of agricultural biomass harvested are directly used to produce human food. The amount of agricultural biomass used as animal feed is strongly dependent on the quantitative relation of livestock to humans and in European rainfed agriculture with a strong reliance on animate power it typically accounted for 30–40 GJ/cap/yr.[3] Firewood for producing heat for households, manufacturing and industrial processes accounts for the rest,[4] amounting commonly to between 15 and 35 GJ/cap/yr, depending on climatic conditions, the relative significance of manufacturing/industrial production and the regional availability of woodlands.[5] The main factors determining the specific level of energy throughput are population density, livestock numbers, climatic conditions and the regional availability of energy resources.

Final and Useful Energy

The energetic metabolism of agrarian systems is characterized by a comparatively low conversion efficiency of primary into final and useful
energy.[6] As every step of energy conversion results in energy losses due to
the second law of thermodynamics, final energy availability must be lower
than primary energy use. In the case of Austria in 1830, a primary energy
input of 73 GJ/cap/yr resulted in the availability of 45 GJ/cap/yr of final
energy and 8.8 GJ/cap/yr of useful energy (Krausmann and Haberl 2002).
Assuming that conversion efficiencies may have been similar in other agrarian systems at that time, final energy availability in typical agrarian societies
may have ranged somewhere between 25 and 45 GJ/cap/yr. This final
energy, in turn, could probably deliver some 5–10 GJ/cap/yr of useful
energy. The low conversion efficiency of approximately 60 per cent from
primary to final and roughly 10 per cent from primary to useful energy is,
at least in part, a result of the low 'conversion efficiency' with which human
and animal bodies transform food and feed into work (Smil 1991). Food
for humans and feed for working animals amounted to about one-quarter
of the total amount of final energy supply in Austria 1830, whereas physical work done by humans and working animals accounted for less than 10
per cent of the amount of useful energy delivered. Provision of heat, both
for space heating and as process heat, required the lion's share of final
energy used (over 70 per cent) and amounted to about 90 per cent of the
useful energy delivered (Krausmann and Haberl 2002). The amount of heat
needed for space heating is, of course, strongly dependent on climate
conditions, and therefore varies greatly from region to region.

Energy and Agriculture

In agrarian societies, agriculture is the most important socioeconomic
activity to provide humans with the required primary energy. Within the
agrarian socioecological regime, it is indispensable that agriculture produces a positive energy return on investment. This means that the amount
of energy invested by society into agriculture or forestry has to be smaller
than the output of energy contained in the agricultural products or wood
gained through these activities (Hall, Cleveland and Kaufmann 1986). If
we consider working animals as an internal element of agricultural production systems,[7] then the only energy invested into the cultivation of the
land (and raising the livestock) is human labour. The food required to
sustain the work invested by humans in cultivating the soil and maintaining soil fertility has to be significantly smaller than the actual food output.
It has been estimated that in European historical rainfed agriculture the

food equivalent of agricultural labour amounted to roughly 15–20 per cent of the total food output of a regional agricultural production system, that is, the relation of energy output to input was 5:1 (Krausmann 2004). In other regions these factors will be different, but they should always be expected to be significantly greater than 1.[8]

Energy and Transport

The area dependency of biomass and its comparatively low energy density (below 15 megajoules per kilogram [MJ/kg] fresh weight for most kinds of biomass, compared with 25–30 MJ/kg for hard coal and 40–45 MJ/kg for petroleum), makes the energy costs of transport a crucial limiting factor. Wood, feed or staple food and other bulk resources cannot be transported over large distances over land, and water-based transport is the only feasible way for long-distance transportation of bulk goods.[9] It is estimated that in general the overland transport range of bulk materials could not exceed 10–50 km (Boserup 1981; Fischer-Kowalski, Krausmann and Smetschka 2004).[10]

Hence, the local availability in relation to local demand and not the average resource/energy availability over a larger territory was decisive. This is illustrated by the energy requirements of pre-industrial iron production: in the 18th century, the production of 1 ton of iron may have required some 50 m^3 of fuelwood equal to the yearly (sustainable) yield of 10–20 ha of woodland. A typical iron smelter with an annual capacity of 500–1000 t/yr, therefore, requires an energy hinterland of 5000–20 000 ha (50–200 km^2) of woodland with an average transport distance for fuelwood of 4–8 km. The energy costs of transport limit the concentration of energy-intensive industrial production and population to certain locales with favourable topographic conditions. Large concentrations of energy consumption must rather be seen as the exception.

Managing Soil Fertility and the Potential of the Land-based Energy System

In the agrarian regime, practically all available primary energy is area-bound biomass. By the 19th century, in Europe almost all productive land was used for energy provision, that is, biomass production for fuel, feed and food. According to a rough estimation, early 19th-century Austria used 40 per cent of its total area for the provision of food, 10–15 per cent for draught power and 30 per cent for heat. Biomass for non-energetic purposes (for example, construction wood) was produced on less than 10 per cent of the land. Biological productivity (that is, aboveground NPP) and the technological possibilities for optimizing the output of socioeconomically useful biomass have a decisive influence on the amount of available energy and result in some typical features of the metabolism of agrarian societies. Before discussing the limitations and potentials of the land-based energy system, we have to take a closer look at the basic characteristics of the functioning of agriculture under the conditions of the controlled solar energy system.

Although market relations and migration have been important elements of local agricultural production systems for centuries, from a sociometabolic perspective local agrarian communities appear to be biophysically more or less closed with respect to socioeconomic flows of materials and energy. External energy and nutrient subsidies are (practically) absent. Basically, agriculture, with respect to the provision of endosomatic energy and the replacement of plant nutrients, exclusively relies on natural cycles and local socioeconomic resources in agrarian societies. This entails a complex optimization of the use of locally available resources that is based on nutrient transfers between different landcover types, closing of internal loops, a minimization of losses and ultimately a dependence on natural flows for nutrient replacement. As a consequence, agricultural production systems are integrated at a farm or village level and characterized by a local mix of cropland, grassland and woodland and the multiple uses of land in agrarian societies. Regional specialization, and thus larger-scale patterns and structures depend on climatic and topographic conditions. Livestock plays a crucial role in the local integration and the optimization of the land-use systems: farm animals provide the required labour, allow for active fertilizer management, supply additional food and raw materials and add value to non-edible crop residues, food wastes and land not suitable for cropping.

An optimal management allows soil fertility to be stabilized and the land to be cultivated with a positive energy yield (see above), albeit at rather low rates of biomass extraction. Yields per area are comparatively low, cereal yields typically ranging between 1–2 t/ha/yr in optimized systems. In order to maintain a viable nutrient balance, it is essential that the internal turnover of biomass is large compared with the exports of biomass: in general only a few per cent of the internal biomass turnover are exported. As a consequence, most of the population live on and from land, and only one-tenth or possibly one-fifth can live in urban agglomerations on occupations other than agriculture. Important elements or technologies in terms of improving the productivity of traditional agriculture during the final phase of the European agrarian regime in the 18th and 19th centuries were the introduction of leguminous crops, reducing nutrient losses due to improved manuring techniques and elaborate irrigation systems. These technologies allowed for a doubling of area yields and a multiplication of food surplus, an important prerequisite of growth and urbanization. However, increasing socioeconomic appropriation of biomass under the conditions of the agrarian regime was possible only within rather strict limits, and exceeding these limits inevitably resulted in soil degradation and declining soil fertility and yields.

A theoretical derivation of the potentials of energy provision and the growth limits of the agrarian energy regime is a difficult task, as it would have to be based on assumptions on the hypothetical scope of improving land use, agricultural technologies and biological productivity under the conditions of the controlled solar energy system. Some insights can be gained, however, through estimates based on empirical knowledge of maximum biomass yields of Central European agriculture in the late 19th century.

In temperate climates, the biological (aboveground) productivity of the natural vegetation averages roughly 200 GJ/ha/yr or approximately 13 t/ha/yr of biomass. The transformation of pristine woodlands with such a high NPP value into managed land-use systems with their local mix of intensively used cropland and extractively used woodlands and grasslands significantly reduces the biological productivity. Only a certain fraction of the NPP is of potential use to human societies and can be appropriated without damaging soil fertility and other ecosystem services in the long run. We assume that in optimized land-use systems under favourable soil and climate conditions and on a larger scale, biomass yields[11] on cropland may reach 3–5 t/ha/yr, on grassland 1–6 t/ha/yr and in woodlands 3–5 m³/ha/yr. If we consider that on larger spatial scales a certain share of the land is of marginal productivity and a certain amount of land is lost for transport, settlement, and so on,[12] we arrive at maximum energy densities of 50 GJ/ha/yr, or at optimistic average values of 30 GJ/ha/yr.[13] A significant increase of biomass yields per area above this level can be achieved only with fossil-fuel-based energy subsidies such as mechanical power, fertilizers and pesticides.[14] On a larger scale an energy yield of 30 GJ/ha/yr may support population densities of 40 persons/km² (see Boserup 1981) resulting in an average per capita energy use of approximately 70 GJ/cap/yr. Increasing the energy yield per unit of area is usually linked to increasing population and hence a tendency of per capita energy use to decline.

Material Use

The inherent limitations of the biomass-based energy system (that is, high energy costs of transport, low energy density, lack of conversion technologies, reliance on endosomatic energy) also shape the patterns of material use: except for biomass, all materials are used at rather low quantities both in terms of volumes per capita and per area. Our reconstructions of the historical metabolism of Austria and the United Kingdom show that domestic material consumption (DMC) ranged somewhere between 5 and 6 t/cap/yr. Also in terms of weight, biomass accounted for the largest fraction of material use, roughly 80 per cent in the United Kingdom and 87 per cent in Austria. We estimate that the use of construction materials amounted to approximately 0.5–1 t/cap/yr. Additionally, a few hundred kg/cap/yr of ores and other minerals such as marl were used. Coal was also used under the conditions of the agrarian regime, but only in comparatively small quantities and at a local level. In Austria, coal use per capita amounted to 5–10 kg/cap/yr around the year 1800; in the United Kingdom the corresponding value was much higher already in the early 18th century, and amounted to 100–300 kg/cap/yr – but this was still less than a few per cent of total domestic material consumption.

8.3   THE AGRARIAN–INDUSTRIAL TRANSITION: A SUMMARY

The transition from the agrarian to the industrial socioecological regime is both a historic and an ongoing process. Table 8.1 summarizes some core indicators used in this book to understand the agrarian–industrial transition. It contains both historical and current examples: the first two columns present the range typically spanned by agrarian and industrial societies. The column LD presents values for least-developed countries, the column DC those for developing countries and EU for the 15 European Union members of 2000. The following five columns present data for Austria (AUT) and the United Kingdom at various points in time.

Currently, the metabolic profile of the 50 'least developed countries' (LD in Table 8.1), which account for 16 per cent of global land area and 11 per cent of global population, displays the ideal type metabolic profile of agrarian regimes we have reconstructed from historical data: average energy use in these countries amounts to only 33 GJ/cap/yr and 13 GJ/ha/yr, biomass being the major fraction of DEC (92 per cent on average), while fossil fuels amount to only 8 per cent. Total material use is as little as 4 t/cap/yr and 1 t/ha/yr. But even the average values for all developing countries in the world (inhabited by about 60 per cent of the world population) come much closer to the agrarian than to the industrial profile as displayed in Table 8.1. As some of the most pressing global sustainability problems – the surging CO_2 concentration of the atmosphere that causes global warming, and rapid land-use change induced by growing demand for biomass, settlement areas and other ecosystem services – are directly related to these flows, these data

Table 8.1   Metabolic profiles of the agrarian and the industrial socioecological regime

		Agrarian Regime**	Industrial Regime	LD 2000	DC 2000	EU15 2000	AUT 1830	AUT 2000	UK 1750	UK 1830	UK 2000
Population density	[cap/km²]	<40	100–300	40	76	116	42	97	30	76	247
Energy use per cap	[GJ/cap/yr]	50–70	150–400	33	64	205	73	197	63	68	189
Energy use per unit area	[GJ/ha/yr]	20–30	200–600	13	49	216	31	191	19	52	468
Biomass	[%]	95–100	10–30	92	50	23	99	29	94	54	12
Fossil fuels	[%]	0–5	60–80	8	50	77	0.5	61	6	46	78
Other	[%]	0–5	0–20	*	*	*	0.5	10	*	*	10
Material use per cap	[t/cap/yr]	2–5	15–25	4.2	6.8	16	5.8	18.1	5.7	6.4	11.6
Material use per unit area	[t/ha/yr]	1–2	20–50	1.3	4.8	18	2.4	17.5	1.7	4.9	28.7

Notes: * Included in numbers for fossil fuels.
** Typical values for temperate climates and rain-fed agriculture.

Sources: Krausmann, Schandl and Schulz (2003); F. Krausmann, V. Gaube, K.-H. Erb and S. Gingrich, unpubl. results; Schandl and Eisenmenger (2006); Sieferle et al. (2006); Weisz et al. (2006).

underline our argument that the agrarian–industrial transition is a major driving force of global unsustainability.

The Metabolic Revolution: From the Agrarian to the Industrial Regime

The transformation of the energy system results in the abolition of the physical limitations that the controlled solar energy system imposed on socioeconomic growth and spatial concentration processes. It leads to absolute and per capita growth of energy use far beyond the limitations of the agrarian regime: the implementation of the fossil-fuel-powered energy system of modern industrial societies increased per capita use of materials and energy by factors of 3–5. However, material and energy use per unit of area increased much more than per capita turnover, because industrialization was, in its early phases, associated with rapid population growth, the so-called 'demographic transition' (Coale 1960).

Physical growth by factors from 15–25 observed if flows are expressed per unit area, instead of per capita values, illustrate the full dimensions of the metabolic revolution induced by fossil energy carriers and the respective technologies. In current industrial societies, material use may amount to 50 t/ha/yr and energy use to 600 GJ/ha/yr, values far beyond comparable flows of materials and energy in ecosystems. From this perspective, the metabolic revolution merely allowed one sustainability problem to be traded for another: even though the agrarian regime, based on renewable energy sources, could be regarded as sustainable in terms of its basic principles, the growth of the European economies could not have been sustained by tapping into solar flows. Shifting towards the exploitation of concentrated stocks of fossil fuel energy made it possible to overcome the agrarian bottleneck. Although this allowed for rapid physical and economic growth, this growth was based on a finite resource base and resulted in a hitherto unknown, and also in the agrarian regime technologically impossible, degree of human impact on ecosystems entailing environmental problems such as acid rain, global warming or deposition of hardly degradable wastes.

As we can see from the historical cases documented in this book, this metabolic revolution – or, in other words, this socioecological transition – required changes in many dimensions of societies' natural relations. These changes appeared in a certain sequence and were to a great degree interdependent. The biophysical starting point of this transition was changes in the agricultural system: technical advancements in agricultural production and land property reforms allow agricultural labour productivity to be increased and hence surplus to be created. This boosts the growth of non-agricultural population and labour force. The development is enhanced by the advent of exploitation of fossil fuels: the use of coal allows the new non-agricultural population to be concentrated in urban centres, where the labour demand of manufacturing and industry is rapidly growing. The social disruptions associated with the social and regional displacement of population contribute to a breakdown of traditional population controls and result in a sharp rise in population growth. Subsequently, available accumulated capital and technology allow the formation of a 'growth engine' based on the technology complex of coal, iron and railroads plus abundantly available human labour beyond agriculture, which further accelerates growth and concentration processes throughout the 19th century. Figure 8.1 shows the relationship between urban population growth and cereal yields from 1600–2000.

In the first half of the 20th century, growth is interrupted by a period of major events: World War I, the world economic crisis and World War II. After World War II, a new phase of biophysical growth combined with unprecedented prosperity is established, driven by an advanced 'growth engine' based on oil, electricity and related technologies and infrastructures such as the internal combustion engine, electrical engines, cars, electrical appliances and an extensive road system (Ayres and van den Bergh 2005). The metabolic revolution now penetrates all aspects of society, leading to a surge in household energy consumption and quality of life. In contrast to the previous growth period, the growth rate of population now declines and population dynamics calms down. From the 1980s onward, many other biophysical indicators such as energy and material consumption also show a slowing rate of growth.

The stages of this transition process are very similar in the cases of both the pioneering United Kingdom and the latecomer Austria, but while the transition required about 400 years in the United Kingdom, Austria accomplished the same process in less than 200 years. When looking at the development of some of the core variables describing the metabolic revolution (Figure 8.2), we can discern almost prototypical patterns of a transition curve in the share of fossil fuels in primary energy use and the share of the agricultural population. This also applies in the case of population density, but the signs of final saturation are less distinct: even though growth rates of population have declined significantly since the 1960s, population (density) is still on the rise, albeit mostly due to migration rather than reproduction. For GDP per capita, the pattern is different: income is still on the rise with high growth rates, which, in combination with a deceleration of growth of material and energy use, indicates relative (but not absolute) dematerialization.

If we look at the relative speed of these transitions, we do not obtain homogeneous results. With regard to the share of agricultural population, we find that Austria in 1830 resembles the United Kingdom in 1600 (that is,

(a)

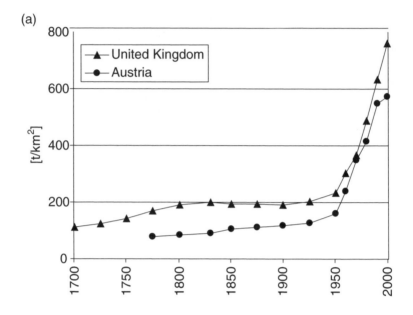

(b)

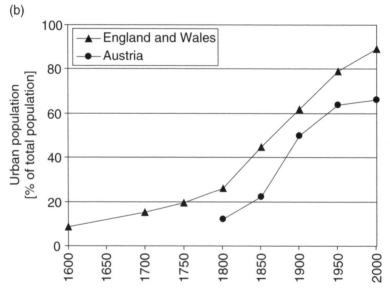

Sources: FAO (2005); Grigg (1980); Sieferle et al. (2006).

*Figure 8.1   Cereal yields a) and urbanization b) in the United Kingdom
and Austria*

a 'time lag' of more than 200 years exists), and that despite a steep decline, it still has not quite 'caught up' with the low UK value by the year 2000. A similar pattern can be found with the share of fossil fuels: again, Austria in 1830 has an equally low (or in fact even lower) share as the United Kingdom in 1600,[15] then despite quickly catching up, in 1910 still lags behind by well over 100 years. The 'time lag' does not decrease in the decades that follow, but instead becomes even larger. This may, on one hand, be due to the effects of the two World Wars, which have hit Austria harder than the United Kingdom, and on the other it may reflect a somewhat different development path: agriculture and renewable resources, above all, wood and hydropower, have played a more prominent role in the Austrian economy throughout the industrial period than in the United Kingdom.

The comparison of population densities tells a similar story, but the difference is even more pronounced: growth rates of population (density) in Austria catch up with the high values of the United Kingdom in the 1840s (when Austria, in terms of population density, lags behind by some 50 years) and throughout the 19th and 20th centuries, population grows at almost the same pace. Consequently, the Austrian population never reaches the UK density and currently is at the level of the United Kingdom in 1880. In terms of income (GDP/capita) too, the gap between Austria and the United Kingdom widened until the early 19th century, when Austria lagged behind by about 120 years. But Austria caught up rapidly thereafter and almost reached the UK level of income before the economic crisis in the 1930s, finally overtaking the United Kingdom in 1970. Summarizing these developments, one obtains the impression that despite the much faster pace of the Austrian transition, it was, according to many biophysical indicators, somewhat softer, causing less environmental harm and overshoot in social and ecological terms.

Two lessons can be learned from this analysis that stand out more clearly because of the very reductionist 'biophysical' approach.[16] Lesson number one is linked to the indicator (primary) domestic energy consumption (DEC), and its uncommon characteristic of both containing the energy needed for the endosomatic metabolism of people and animals, food, and feed, as well as the so-called 'technical' energy requirements that are usually recorded. By looking at the energy base of societies in this broader sense we were able to identify more clearly one of the energetic bases of the transition: the increase in food and feed production that allowed the widespread use of labour power outside of agriculture.[17] We owe the second lesson essentially to Sieferle's (2001) efforts to demonstrate the key role that fossil fuels (coal) played in overcoming the constraints of the traditional solar-based regime, constraints that could never have been overcome by technological innovations based on human ingenuity alone.

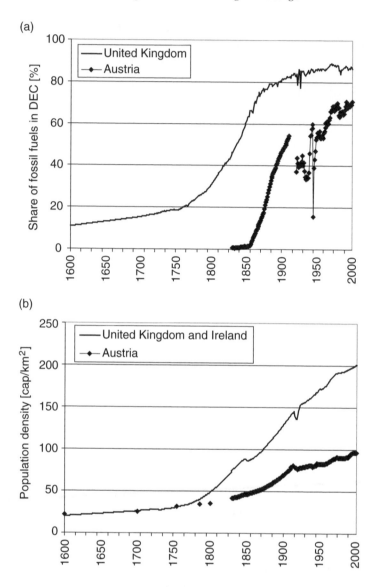

Sources: Figure 8.2a, b and c: own calculations on the basis of data presented in Chapters 2 and 4; Figure 8.2d: Maddison (2001).

Figure 8.2   Two historical transitions compared: the United Kingdom and Austria in terms of a) share of fossil fuels in primary energy use, b) share of agricultural population, c) population density and d) income (GDP/capita)

(c)

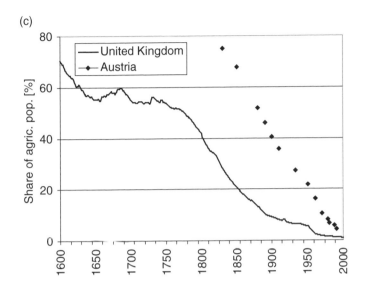

(d)

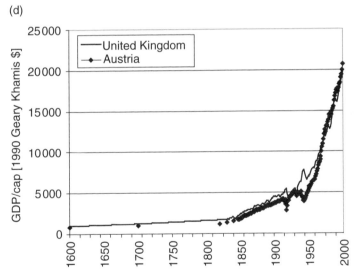

8.4 CAN LESSONS BE DRAWN FROM PAST TRANSITIONS FOR CURRENT ONES?

The comparative perspective looking at the United Kingdom, where the transformation started, and Austria, which was a latecomer (at least by European standards), teaches us that socioecologically, these two countries were rather alike in 1600: both countries had approximately the same population density, similar agricultural yields, a similarly high share of agricultural population, similar energy consumption and – as far as this can be reliably compared – similar income. Transition in the United Kingdom started to take off in the middle of the 17th century, first on the basis of its own agricultural surplus owed to agrarian reform, then moving towards the use of coal, iron and railroads. It stayed on this high energy/heavy industry track for 300 years until well into the 1970s, when it finally underwent a period of structural adjustment and de-industrialization. Austria underwent its take-off in the middle of the 19th century, was for a long time far behind the United Kingdom in terms of all indicators of industrial development, and finally caught up regarding standard of living in the 1970s, with somewhat less severe social and environmental costs than those seen in the United Kingdom. Austria as a latecomer, benefiting from technological and economic advances, managed the same transition process within a period 200 years shorter than that required in the United Kingdom.

Can we expect something similar to happen where contemporary developing countries are concerned? Where do they stand in relation to such a transition, and which socioecological 'catching up' patterns can we discern? Alternatively, might there be diverging development paths? Based on the data discussed in the preceding chapters, we use a new analytical tool: we compare key indicators of transition for the developing countries in the year 2000 with the United Kingdom across time. We express what we see as 'years' time difference' (Müller and Sicherl 2004). Thus, we take into account the speed with which the United Kingdom changed this parameter in its transition: small differences in values may take a very long time to overcome, and sometimes large differences may happen very fast. This method, of course, only makes sense with parameters that increase or decrease steadily over time.[18]

We can use the information presented in Table 8.2 for two purposes. First, to understand whether we are indeed dealing with the same process as happened in the past, or whether a different process is taking place. Second, is there one pattern of transition for contemporary developing countries, or are there several patterns? A positive answer to the first issue would imply a certain homogeneity of time distances: the 'delay' in population density should resemble the delay in agricultural population, energy use and income. This is clearly not the case: rather surprisingly, in most cases, the time difference in income is smallest. With the notable exception of Laos and Vietnam,[19] there is a maximum time lag of 150 years. At the other extreme, the share of agricultural population in some countries resembles that of the United Kingdom more than 300 years ago.

Table 8.2   *Time lag in the transition process: developing countries in 2000*
relative to the United Kingdom (values in the cells represent the
approximate number of years' time difference since the United
Kingdom had the same parameter value)

	Population Density (cap/km²)	Share of Agricultural Population (%)	Share of Fossil Fuels (% of DEC)	Income (GDP/cap in 1990 Geary Khamis $)
Brazil	−400	−130	−180	−60
Chile	−400	−130	−150	−40
Venezuela	−400	−110	−140	−40
Laos	−400	−400	−400	−400
Philippines	+20	−200	−170	−150
Thailand	−100	−220	−150	−70
Vietnam	−10	−400	−220	−400

Sources: See Figures 8.1 and 8.2 for the United Kingdom, and Appendix 7.1, Chapter 7 for the developing countries; based mainly on Maddison (2001).

What we can also see is a fairly close correspondence between the share of agricultural population and the share of fossil fuels in terms of time distance. This confirms our theoretical explanation that beyond a certain threshold, the use of other energy sources is instrumental in liberating people from agricultural work. The most inconsistent results appear with reference to population density: while population density in the Latin American countries corresponds roughly to the United Kingdom in 1600, it is equal to or even higher than the United Kingdom now in some Southeast Asian countries. On the question of whether this contradicts our theoretical model, we feel it does not, on the contrary, and we discuss this further below. Before doing so, we try to provide an answer to the first question. We can see that in contemporary developing countries there is a pattern of transition similar to the (European) past, at least in the sense that all parameters in developing countries correspond to the United Kingdom's past rather than its present. So the directionality of processes is as should be expected. Nevertheless, the homogeneity of time lags is very low indeed.

Now to the second question: is there broadly speaking one socioecological pattern, or do several patterns of contemporary transitions exist? Here the answer seems very clear to us, and it corresponds to Boserup's theory of agrarian development (Boserup 1965; 1981). In these profiles, we can discern two patterns, an 'Old World' and a 'New World' pattern. The point of agrarian development at which the transition to the industrial mode occurs, that is, how far the agrarian regime has already been driven to the limits, is decisive. With regard to this difference, the most important distinction is between Eurasia (the 'Old World'), on the one hand, with an agrarian development spanning 5000–10 000 years since the onset of the Neolithic revolution, and America and Australia (the 'New World') on the other hand, with a few centuries of agrarian history[20] and correspondingly low population densities. These two 'Worlds' are marked by two different trajectories:[21] the New World trajectory is much more expansive, resource-intensive and 'generous', due to the fact that a pioneer situation still exists, an open frontier to an as yet uncolonized nature. In the Old World, in contrast, natural resources have already been very much stretched to the limits under the agrarian socioecological regime.

There is also a specific difference within the Old World pattern, between (historical) Europe and Asia. One of the specific aspects of the Asian situation and its transition pattern, in comparison to, let us say, 17th-century England, is that Southeast Asian countries, in accordance with Boserup's theory, have proceeded much further along the path of an agrarian regime that is both land- and labour-intensive. With traditional methods (that is, without fossil fuel for irrigation, machines and chemical fertilizers) they were and still are unable to achieve the further improvement in labour productivity on which European countries in history built their 'great transformation'. While Europe, with the help of coal, achieved a take-off at a time when it still had the chance to improve substantially on agriculture with traditional methods, Southeast Asia proceeded on the agrarian path and drove it much further to the limits, so that this use of traditional methods in combination with population pressure[22] makes it difficult to use (and abuse) agriculture for revenue to be invested in industrial development.

On the basis of the analysis presented here, it is not particularly easy to identify the impact that the colonial history may have had. In Chapter 4 we were able to identify some of the benefits reaped by the United Kingdom historically from its colonies (benefits that neither historical Austria nor any of the developing countries could draw upon). But it is not so clear how long-lasting these benefits are, in terms of the variables we focus on. Conversely, we may ask how enduring the negative impacts of colonialism on the former colonies have proved to be, and what they consist of within the range of parameters considered. The most plausible hypothesis would be that colonial domination and wars prolonged the agrarian socioecological regime, and delayed the onset of industrialization. What it did not achieve, as far as can be judged from the data presented here, was that the developing countries' income would lag further 'behind' the industrial core than their structural characteristics – indeed, quite the opposite is the case.

8.5   THE BOOK'S FOUR KEY QUESTIONS – AND A GLIMPSE OF THE FUTURE

In the introduction to this book, we have raised four overarching questions. We will now return to them and try to articulate some of the answers that we feel were given in the intervening chapters.

Question 1:   Is there such a thing as a characteristic metabolic profile of agrarian societies?
The simplest answer is: yes, a characteristic profile does exist and it is qualitatively and quantitatively clearly distinct, in terms of energy and materials input per capita per year, from an industrial profile. A more sophisticated answer is that this was not exactly the right question. Its formulation suggests that agrarian systems exist in a static state. What we did find, both by time-series analysis and by interpreting snapshot data accordingly, were trajectories of development of agrarian systems. Contrary to widespread assumption, but very much in accordance with some of the theoretically more sound literature (such as Ester Boserup's writings), we found the agrarian socioecological regime to be quite dynamic.

One of the major factors underlying this dynamic characteristic seems to be population-growth-driving innovations in agricultural production practices and technologies. These production innovations lead to an increase in area productivity, often at the expense of labour productivity (Boserup 1965; Netting 1993; Sieferle et al. 2006). Institutional changes (such as property reforms creating larger and more rationally managed estates) may lead to a reduction of excess labour force in agriculture (increasing the pressure on the remaining labour force) and allow for a higher surplus of labour and agricultural produce to be invested in non-agricultural activities. This presumes the existence of a transport infrastructure (harbours, roads) that allows bulk goods to be moved cheaply over larger distances, the existence of markets, and, in the long run at least, the availability of gainful mass employment outside agriculture.[23] While we may believe that this point in the development of agrarian regimes has been reached at many times and in many places, it was only in the singular case of the United Kingdom that a new energy source was both readily available[24] and represented an opportunity that could not be turned down for lack of alternatives.[25]

Question 2:   What happens when this socioecological regime starts to change? What are the changes both in the social and the natural systems concerned?
The findings from the case studies assembled in this book make us believe that transitions follow a certain common formal pattern. The methodological

task in analysing transitions would then be to identify this pattern and to name and quantify the relevant parameters. Formally, the pattern of a transition can be described as follows.

A transition begins with some form of 'steady state', a dynamic equilibrium or (relative) balance. This dynamic equilibrium can be characterized by processes in which elements that have some in-built dynamics or that are linked to one another by positive feedback loops are counterbalanced by negative feedbacks that prevent change beyond a certain point. As all historical situations are, of course, over-determined, the art of formulating the conceptual model consists in selecting just a few core elements and processes that 'do the job' of keeping the system in balance.

At some point in time, one or more negative feedbacks built into the 'steady state' are removed, and this unleashes a transition process, that is, a sudden growth or decline of one or more critical variables beyond the range to which they had been restricted before. This characterizes the 'take-off' phase of a transition. A dynamic system in its 'acceleration phase' of a transition is not only characterized by the removal of one or several negative feedback loops that prevent change, but seems to be driven by some kind of 'growth engine', that is, a closed cycle of interlinked positive feedbacks, that drive the system towards states it had never been in. This phase may not only be nourished by one such dynamic process, but by a sequence or an interlinkage of several such processes that trigger or reinforce one another but can clearly be distinguished.[26] Finally, a transition grinds to a halt not only due to the exhaustion of the growth engine, but also through the establishment of one or several new negative feedback processes that keep it in a new state of balance.

Analysing transitions means identifying and describing these positive and negative feedback loops and understanding the growth engine(s) at work. Following these methodological guidelines in analysing a transition goes far beyond identifying trends in a few variables and characterizing their statistical properties. The ultimate test of whether the key properties and their interlinkages have been identified would, of course, be to create a formal dynamic systems model, which is still rather far from where we are now. What we are very close to, however, and have approached with many of the case studies in this book, is the development of a consistent conceptual model that would allow us to analyse these different cases according to a common latent pattern. As long as this has not yet undergone rigorous testing, however, we still consider it to be fairly speculative, especially in view of the small number of variables we were able to take into account. Here we expect to provoke some irritation, particularly on the part of historians and social scientists, who are used to pointing to the complexity of each and every real situation, and we certainly would not claim that this approach does not make sense, or could be substituted by a reductionist modelling approach such as the one we present. Our goal is a different one: we wish to identify a particular socioeconomic and socioecological process that has taken place and is taking place many times, on many scales and under varying contextual conditions all around the world, and characterize it in a way as simple as possible to still be able to identify each particular case as being at a certain stage and involved in a particular dynamics outlined in the general pattern.

By describing transitions the way we do it, we do not wish to imply that they follow any closed or inevitable sequence of events. On the one hand, in the course of any transition there are various points where, according to our conceptual model, exogenous factors or framework conditions come into play that make a difference. On the other hand, all variables in complex systems can be assumed to relate to one another in a non-deterministic, stochastic fashion, and therefore there are always opportunities for a different course of events to occur. And finally, there are so many relevant variables that we leave out of our model altogether that we should not be surprised to find that very deviant cases in fact occur.

Question 3:   How does the transition from the agrarian to the industrial mode depend on the world context?
The historical comparison of the United Kingdom (pioneer) case and the Austrian (latecomer) case teaches us that latecomers may undergo the agrarian–industrial transition and acquire similar economic prosperity in a much shorter time period and with less social and ecological hardships than in the case of the pioneering Britons. But does this hold equally for contemporary developing countries? One apparent difference is that they base their industrial transformation less on the exploitation of an internal agricultural surplus,[27] and more on an international financial market. The implications of this are widely discussed, and the preliminary conclusions from that discussion are certainly ambiguous (see, for example, Stiglitz 2002).

Production structures and products, not only in terms of investments but also in terms of raw materials, are currently much more dependent on a world market dominated by the rich industrial core countries. Among the countries we analysed we found two quite distinct patterns. One group of countries relies on the extraction of natural resources, thus becoming 'extractive economies' (Bunker 1984). This pattern seems to be prevalent in countries that, due to a relatively short history of agrarian settlement, have a low population density but are well endowed with natural resources such as mineral resources or productive land. A second group of countries, those with a high population density, develop a different pattern that relies on their capacity to supply cheap semi-skilled or skilled labour: the 'global sweatshop' (Schor 2005). In both cases, production is export-oriented and products are consumed elsewhere, while the social and environmental costs of production and the depletion of resources are felt at home.

There is another contextual feature that has been relevant in the past and may become even more crucial in the future: the availability of energy and other mineral resources at a fairly low price. It has been shown that periods of rapid growth of the industrial core countries were mostly also periods of decreasing (relative) energy prices (see, for example, Fouquet and Pearson 1998; Pfister 2003). A large literature exists, moreover, that has demonstrated how much importance was attached to gaining and securing easy and cheap access to raw materials and other biophysical resources in the history of European colonialism (Bunker 1996; Bunker and Ciccantell 2003; Hornborg 2005) – in particular to those raw materials that were required to support the industrial transition. Such a world context, and such strategies, are not any more available for contemporary developing countries. Their transition phase coincides with energy prices that are rising at present, and are highly likely to continue to rise in the future. Their transition takes place in a period in which the scarcity of at least some raw materials can already be felt, even if this may not yet be reflected in their current price levels. These contextual variables have unfortunately not been subject to our analysis, but we will use them when we consider scenarios for the future.

Question 4:   How does the interplay between different levels of functional integration and spatial scales work, and what role do scale interactions play in socioecological transitions?

The case studies presented in Chapters 5 and 6 suggest that the agrarian–industrial transition implies fundamental changes in rural–urban relations and thus also in interactions between spatial scales. Under the agrarian regime, the transport of large amounts of material is extremely expensive and therefore most material and energy flows are local. Transport of significant amounts of material over larger distances can only be accomplished on waterways, be they rivers, lakes or the ocean. Local socioecological systems are therefore largely self-contained and self-sufficient. They receive little, if any, material and energy inputs from other locales and they are able to produce only a small surplus that can be used to support larger settlements or even cities with a significant proportion of non-agricultural population.

This has important consequences for social as well as ecological patterns. With respect to socioeconomic structures it implies, above all, severe restrictions to the division of labour. About 80–90 per cent of the population have to work in agriculture and forestry to generate a surplus that supplies the remaining 10–20 per cent of the non-agricultural population with food, feed and other raw materials (for example, fuelwood, construction wood, raw materials for textiles and so forth). This limits the growth of all non-agricultural sectors such as mining, manufacture and services. Moreover, this restriction also limits the size of cities in relation to the agrarian 'hinterland' required to supply sufficient resources to the city (Fischer-Kowalski et al. 2004; Sieferle et al. 2006). For rural systems, this implies that in each locale there has to be a mixture of cropland, grassland and forests, because each of these three basic land uses is required to supply certain essential and, to a large extent, not substitutable resources, food for humans in the case of cropland, feed for livestock in the case of grasslands and forests, and wood in the case of forests. This meant that cultural landscapes of agrarian societies were diverse and finely structured (Sieferle 1995) and characterized by largely closed local material cycles (Krausmann et al. 2003) – but it also meant that much land was used for purposes it was not, in an ecological and economic sense, perfectly suited to.

Already very early on in the transition to the industrial mode, these tight limits to mass transport were eliminated, first by canals and railway systems and later by fossil-fuel-powered water transport and automotive land transport technologies (Grübler 1998). This, together with abundant energy availability and the spectacular increase in agricultural labour efficiency discussed above, removed these limits and resulted in a complete spatial rearrangement in socioecological systems. At present, transport volumes within nation states as well as trade volumes (in physical terms) between nations are surging. In the industrial world, most of the population today lives in urban centres and agricultural population has fallen to record lows of less than 5 per cent of the total population, thus allowing for a spectacular increase in specialization and division of labour. Limits to urbanization have essentially ceased to exist, and large-scale movement of resources, goods, people and information are being organized. Costs of transport of materials, people and information are falling and transport velocity is increasing, thus rapidly transforming spatial patterns in social organization as well as resulting in large-scale movements of materials (including carbon and plant nutrients such as nitrogen, phosphorous, potassium and so on) and organisms.

As a consequence, globalization is not only a socioeconomic phenomenon, but also a biophysical one. An analysis of any one locality, any one region in the world, has to take into account this fundamental material 'openness' of each socioecological system. Functionally, each locality and each process is co-determined by its role in an increasingly closely integrated global network of interdependencies. This is not entirely new – but never in history has this global network been as pervasive as it is today.

A Glimpse of the Future

How will the transitions we have been describing in this book continue in the future? What possible scenarios does the socioecological perspective we have presented suggest? Does it result in appraisals of the plausibility or implausibility of future trajectories that are any different from those derived from other approaches? We think this is the case, and we will briefly sketch our ideas. In particular, our empirical analyses suggest some fundamental distinctions with respect to transition dynamics that result in a classification of countries into three large categories, first, the industrial core, second, low density developing countries, and third, high density developing countries.

Our first category, the industrial core, includes those countries that have already undergone the transition to the industrial regime or are at least in its very late stages. Within this group we distinguish between 'Old World' and 'New World' countries, as the latter tend to have a much lower population density and a much higher consumption level of natural resources (Table 8.3). Whatever differences the countries assembled in these two subcategories might have, they all have a high level of energy and materials consumption – both directly, and even more so, indirectly – a high per capita income and high technical efficiencies in many important production processes. A third subcategory of the industrial core is the countries of the former Soviet Union and its satellites that also have a high level of material and energy use. Their population density is mostly low (with the exception of the East European countries). They differ from the first two subcategories with respect to per capita income and efficiency, which are both much lower than in the other countries of the industrial core. Together, the countries of the industrial core currently comprise about 23 per cent of the world population (Table 8.3), a value that will drop to about 16 per cent in the year 2050 according to UN population forecasts.

The second category is the low population density developing countries, roughly equally split between New World countries and African countries in terms of population. They are in an earlier stage of the transition to the industrial regime. At least the New World countries in this category already have a comparably high level of energy and materials use, although their per capita income is still much lower than that of the industrial core. They mainly accomplish their further advance in the transition towards industrial society through expansion of their use of natural resources, both biogenic and mineral. They currently comprise about 15 per cent of the world population, a percentage that is expected to grow significantly in the next half century (20 per cent).

Table 8.3 Metabolic profiles of world regions according to development status and population density*
for the year 2000

	Share of Global Population (%)	Share of Global Land Area (%)	Population Density (cap/km²)	Income (GDP/cap) (US$/cap/yr)**	Electricity Use (GJ/cap/yr)	Energy Use (DEC/cap) (GJ/cap/yr)	Energy Use (DEC/area) (GJ/ha/yr)	Material Use (DMC/cap) (t/cap/yr)	Material Use (DMC/area) (t/ha/yr)
Industrial core (Old World)	13	5	123	18048	22	193	238	14	17
Industrial core (New World)	6	21	12	27404	52	443	54	30	4
Industrial core (Former USSR)	5	17	13	1740	16	176	23	14	2
Low population density developing c. (New World)	6	14	19	2831	7	130	25	13	2
Low population density developing c. (Africa/Asia)	9	24	17	1384	4	76	13	6	1
High population density developing c.	62	20	140	810	3	49	69	6	8
World	100	100	45	4665	9	102	46	9	4

Notes: * Low population density: countries with a population density < = 50 cap/km², high population density: countries with a population density > 50 cap/km².
** GDP in constant prices, 1990 US$.

Sources: Own calculations based on FAO (2005) (population); UN Statistics Division (2004) (income); IEA (2004) and UN (2004) (electricity use); F. Krausmann, V. Gaube, K.-H. Erb and S. Gingrich, unpubl. results; Schandl and Eisenmenger (2006); Weisz et al. (2006) (energy and material use).

Our third category is the high population density developing countries, most of which are 'Old World' Asian countries (85 per cent of the population in this category). They are often in an even earlier stage of the transition, still have a large agricultural sector, a low level of per capita energy and materials use, low income and a large and often well-qualified labour force upon which they base their advance in the industrial transition. They comprise about 60 per cent of the world population now, and an even larger proportion in the future.

What would be the crucial variable, the variable that would make the largest difference in terms of influencing the transition pathways of these countries? For the agrarian regime, this would clearly have been the size times the productivity of their territory – a weighted measure of area. Under present conditions, in the transition to the industrial regime, we think that the (relative) price of energy will play a crucial role, and we will qualitatively distinguish between *high energy price* scenarios and *low energy price* scenarios, without choosing any explicit quantitative dividing line.

Let us now engage in a thought experiment on the fate of our three categories of countries under these different assumptions. For the low energy price scenario, we would expect a continuation of current trends for quite a while. The industrial core would be able to stabilize its energy and materials consumption at a high level and would probably continue to enjoy moderate economic growth rates. We would expect the low population density developing countries to experience an explosive rise in the extraction and use of resources, in particular we would expect surging energy consumption there. These countries would continue to supply raw materials to a rapidly growing world demand, and to some extent they could benefit economically from this trajectory. The high population density developing countries would continue with a rapid, labour-intensive economic growth, and could suffer from exploding pollution problems. These countries, as we have seen before, may be characterized by low energy and materials intensity on a per capita basis, but their energy and materials intensity per unit area is already now very high due to their large population density (Table 8.3). Further increases in per capita rates of resource use together with a continuation of population growth are bound to raise local and regional pollution levels further. In the absence of vigorous pollution abatement measures, these problems might even aggravate to devastating levels. Globally, this low energy price scenario can be expected to result in a rapid depletion of resources, rapid global environmental change and, in particular, rapid climate change.

In the case of the high energy price scenario we would expect the industrial core to reduce its energy and materials consumption, utilizing efficiency, savings and externalization potentials, and maybe struggling with a lack of economic growth. The low population density developing countries would probably fare best because they could use their endowment with large productive land areas to produce biomass for energy generation for their own consumption and probably also for export to other regions. These countries would also continue to exploit their mineral resources, maybe at a lower rate, while building up their own industrial capacities. The high population density developing countries, hosting more than half of the world population, could get into real trouble in this scenario. Scarcity of affordable energy and mineral resources could choke their economic growth and result in severe political conflicts over resources and social distribution. These conflicts could even escalate to global military threats. So this second scenario, while less bleak than the first in environmental terms, might turn out to be catastrophic in terms of political stability and social equity.

Are there alternatives to Scylla and Charybdis? There does not seem a 'middle of the road' solution: medium energy prices rather combine the risks of both alternatives in a dirty mix. Nor does a 'wait and see' attitude appear to be very promising. It may well be true that information technologies (IT) are less energy demanding than conventional industrial technologies, yet they depend on minerals, some of which are extremely rare. Moroever, IT is implemented in addition to – and not as an alternative to – resource-intensive sectors such as housing, transport and food supply. It may allow for a continuation of less resource-intensive growth ('relative decoupling') in industrialized countries that have already built up the resource-intensive supply systems for these demand categories, but it doesn't offer an alternative development path for countries that are in an early stage of industrialization. While rising oil prices will of course pave the way for alternative energy sources, the industrial core is locked in with a precarious infrastructure in terms of transport, energy and water supply, housing and so on, which requires a continuous supply of high amounts of energy and materials each year and can only gradually be changed. We believe that finding a path towards more sustainable solutions requires a vision and determined political efforts that include at least the following items:

- A renewed global commitment to climate policy that leads to a stabilization and subsequently to a reduction of fossil fuel combustion before this is imposed by resource scarcity. This requires the implementation of a new kind of industrial energy system based on new, efficient technologies that might look very different in each world region, depending on its characteristics in terms of population density, climate and so on. Both energy conservation and new renewables, including solar heat and power, geothermal and wind energy, hydropower and a cascade utilization of biomass may play a role here, although certainly none of these technologies alone will do the trick.

- A process of learning to accept self-restriction in terms of resource use in the industrial core, and learning to resolve distributional problems other than with rapid economic growth. This need not be so unalluring. One could even argue that more freedom in disposing over one's time, and improved embeddedness in meaningful social relations could be more important for one's quality of life than increased consumption of material goods.
- A concerted effort at inventing, designing, developing and testing new types of infrastructure that would not lock in the rest of the world in a pattern that requires large yearly flows of energy and materials. The most urgent need for this initiative certainly exists in the high population density developing countries, and these may eventually have the collective sense of responsibility plus the required decision-making capacity lost in much of the industrial core to actually succeed with such an endeavour.
- A global responsibility for and measures to prevent the last areas of pristine wilderness from final destruction. Additionally, measures are needed to maintain and foster biodiversity in human-dominated areas also, so that the world heritage of biological evolution be preserved for future generations.

These may seem challenging, even illusionary demands upon humankind's limited capacity to shape its future in a purposive way. But we hope that this book has been successful in conveying the message that our past trajectory has not been very likely either. This book also suggests that there are futures that are impossible – but the range of possible futures is very broad indeed, and it seems worthwhile to make a serious effort at hitting some of the brighter areas.

NOTES

1. Chapters 2, 4, 5 and Haberl et al. (2006); Krausmann and Haberl (2002); Weisz et al. (2001).
2. Note that these energy sources were of vital importance for the functioning of the economy, though, by providing crucial inputs of mechanical power used for transport and important production processes (for example, mills). See Smil (1991, 1994).
3. In European land-use systems, livestock plays a crucial role. Agriculture depends on livestock in several ways: draught power, manure, food and raw material production. In many land-use systems of agrarian societies, therefore, more than 90 per cent of all agricultural biomass is used for livestock (feed, litter, grazing). See Krausmann (2004).

4. On a regional level and in some countries, peat may play an important role in energy provision.
5. Depending on the significance of energy-consuming manufacturing/industry, the share of households in total fuelwood consumption was probably 60–90 per cent.
6. In these calculations, it is not solar energy itself that counts as primary energy but the solar energy stored as chemical energy in green plants. The conversion of solar energy into plant energy through photosynthesis is also not very efficient. Even under favourable conditions, the net primary production (NPP) of plants – that is, the energy available in chemically stored form for all food chains – is less than 1 per cent of the amount of available solar energy (Odum 1959).
7. The 'fuel' to run draught animals was produced within the farm or region and their draught power was invested within the farm or region.
8. Note, however, that different versions (that is, in terms of system boundaries) of the quantification of the energetic return of agriculture may lead to significant quantitative differences of the results and only data stemming from methodologically consistent estimates can be compared.
9. Access to oceans, favourable river networks and terrains were decisive factors determining the energy costs of long-distance transport and spatial patterns of energy use in agrarian societies (Bagwell 1974; Sieferle et al. 2004).
10. It has been estimated that in the 18th century the transport of wood over 1 km increased the price by 40 per cent in case of land transport, 10 per cent in case of canal transport and 4 per cent over sea (Sieferle et al. 2004).
11. Biomass yields refer to primary products (for instance, cereals) and socioeconomically useful by-products (straw). Values are given in air dry weight, that is at 16 per cent water content, conversion into energy values was made assuming a calorific value of 15 kg air dry weight.
12. For instance, in the United Kingdom and Austria, roughly 10–20 per cent of the land was unproductive barren lands or settlement areas and another 10–20 per cent were areas of marginal productivity only useful for extensive grazing.
13. Optimized refers to land-use systems relying on crop rotation without fallow, the use of fodder legumes and sustainable forest management as, for example, in 19th-century United Kingdom or late 19th-century Austria.
14. Industrialized land-use systems in Europe yield 100–200 GJ/ha/yr.
15. Sieferle (2001, pp. 86ff and 104) gives figures for the United Kingdom's coal use in the period 1551–60 amounting to 200 000 t annually. This corresponds to 5.6 PJ/yr (about 1 GJ/capita/year). We may assume a primary energy consumption of about 60 GJ/cap/yr; thus coal accounts for 1–1.5 per cent. This is not as little as it seems: the amount would suffice to substitute one-fifth to one-half of the fuelwood in the urban households of the time.
16. We have hardly touched upon any of the social or political repercussions of these historical transitions. By disregarding them in this way, we certainly do not wish to imply their irrelevance. As touched upon in Chapter 1, we expect the take-off phase of a transition to be marked by social and political conflict eventually facilitating the technological and biophysical changes we describe. Thus, the 'Cromwell revolution' in England in 1588 marks the United Kingdom's take-off at a critical point in time, similar to the revolutionary activities in Austria in 1830 and 1845. Beyond that, if one looks at the timeline of other European revolutions, it appears very much in tune with the timeline of other take-offs. However, it certainly transcends the scope of this book to offer any narrative that would systematically link biophysical and social transitions.
17. There is, of course, a social, political and cultural background to this that we barely mention: religious conflicts and liberations that impact on landed property, its use and exploitation, and the inspirations drawn from 'enlightenment' concerning more rational resource use and technology development. We wish in this case to complement the arguments more commonly brought forward to explain the Industrial Revolution, not to detract from their validity.

18. In cases where the United Kingdom passed through the same parameter value more than once (as a consequence of disturbances during the World Wars, for example), we used the earliest point in time when a certain value was reached as reference time for the comparison.
19. As we were able to see with the historical cases of the United Kingdom and Austria, wars cause major disruptions of the parameters observed, particularly concerning income. It seems likely that several of the special features of Laos and Vietnam may be traced back to the impact of the long-lasting Indochina and Vietnam Wars.
20. Indigenous agrarian traditions reach back further, but they covered only small portions of the countries with much 'uncolonized' space in between.
21. There are of course not only these two 'Worlds', but also Africa and the former Soviet Union would be expected to display their own specific patterns. However, we know too little about these regions to incorporate them in this concluding chapter.
22. Oesterdieckhoff (2000) supplies an interesting hypothesis according to which different family structures between Western Europe and Asia have played an important role in economic transformation. The European 'collateral' pattern links the chance to marry and have children to the economic success of the sons (they must be able to sustain a family), while the patrilinear joint families characteristic for Asia do not relate the chance to have children to the son's economic independence and so allow for population growth independent of conditions of prosperity. This favours a course of history where population growth always 'eats up' productivity gains and indeed often exceeds these gains, leading to crises of starvation, civil strife and epidemics.
23. This may have been one of the challenges facing the Roman Empire: unless there was an army engaging the surplus population, and feeding it (at least partly) at the expense of defeated enemies, the jobless plebeians became a threat to political stability.
24. Diamond (2005); Pomeranz (2000); Sieferle (2001).
25. At that time, there was hardly any forest left over in the proximity of population centres, so coal, whether it spread an unpleasant odour or not, had to be utilized (Sieferle 2001).
26. This idea, of course, comes very close to the theory of 'Kondratieff Cycles' (Wallerstein 2000) in technological development. We, however, believe a larger range of variables to be involved in the dynamics of each phase.
27. As was the case both with the United Kingdom and Austria, and probably also with much later cases such as Japan in the last quarter of the 19th century and the Soviet Union after World War I.

REFERENCES

Ayres, Robert U. and Jeroen C.J.M. van den Bergh (2005), 'A Theory of Economic Growth with Material/Energy Resources and Dematerialization: Interaction of Three Growth Mechanisms', *Ecological Economics*, **55**, 96–118.
Bagwell, Philip S. (1974), *The Transport Revolution from 1770*, London: B.T. Batsford.
Bell, Daniel (1973), *The Coming of Post-industrial Society: A Venture in Social Forecasting*, New York, NY: Basic Books.
Boserup, Ester (1965), *The Conditions of Agricultural Growth. The Economics of Agrarian Change Under Population Pressure*, Chicago: Aldine/Earthscan.
Boserup, Ester (1981), *Population and Technological Change – A Study of Long-term Trends*, Chicago: The University of Chicago Press.
Bunker, Stephen G. (1984), 'Modes of Extraction, Unequal Exchange, and the Progressive Underdevelopment of an Extreme Periphery: The Brazilian Amazon, 1600–1980', *American Journal of Sociology*, **89**(5), 1017–64.

Bunker, Stephen G. (1996), 'Raw Material and the Global Economy: Oversights and Distortions in Industrial Ecology', *Society and Natural Resources*, **9**(4), 419–29.

Bunker, Stephen G. and Paul S. Ciccantell (2003), 'Creating Hegemony via Raw Material Access. Strategies in Holland and Japan', *Review. A Journal of the Fernand Braudel Center for the Study of Economic, Historical Systems, and Civilizations*, **26**(4), 339–80.

Coale, Ansley J. (1960), *Demographic and Economic Change in Developed Countries*, Princeton, NJ: Princeton University Press.

Diamond, Jared (2005), *Collapse. How Societies Choose to Fail or Succeed*, New York: Viking.

FAO (2004), *FAOSTAT 2004*, FAO Statistical Databases: Agriculture, Fisheries, Forestry, Nutrition, Rome: FAO.

FAO (2005), *FAOSTAT 2005*, FAO Statistical Databases: Agriculture, Fisheries, Forestry, Nutrition, Rome: FAO.

Fischer-Kowalski, Marina, Fridolin Krausmann and Barbara Smetschka (2004), 'Modelling Scenarios of Transport Across History From a Socio-metabolic Perspective', *Review. Fernand Braudel Center*, **27**(4), 307–42.

Fouquet, Roger and P.J.G. Pearson (1998), 'A Thousand Years of Energy Use in the United Kingdom', *The Energy Journal*, **19**(4), 1–41.

Galbraith, John K. (1958), *The Affluent Society*, London: Hamilton.

Grigg, David B. (1980), *Population Growth and Agrarian Change. An Historical Perspective*, Cambridge, London, New York, New Rochelle, Melbourne, Sydney: Cambridge University Press.

Grübler, Arnulf (1998), *Technology and Global Change*, Cambridge: Cambridge University Press.

Haberl, Helmut, Helga Weisz, Christof Amann, Alberte Bondeau, Nina Eisenmenger, Karl-Heinz Erb, Marina Fischer-Kowalski and Fridolin Krausmann (2006), 'The Energetic Metabolism of the EU-15 and the USA. Decadal Energy Input Time-series with an Emphasis on Biomass', *Journal of Industrial Ecology*, **10**(4).

Hall, Charles A.S., Cutler J. Cleveland and Robert K. Kaufmann (eds) (1986), *Energy and Resource Quality, The Ecology of the Economic Process*, New York: Wiley Interscience.

Hornborg, Alf (2005), 'Footprints in the Cotton Fields: The Industrial Revolution as Time–Space Appropriation and Environmental Load Displacement', *Ecological Economics*, **59**(1), 74–81.

IEA (2004), *Energy Statistics of OECD and Non-OECD countries* (CDROM version), Paris: International Energy Agency.

Krausmann, Fridolin (2004), 'Milk, Manure and Muscular Power. Livestock and the Industrialization of Agriculture', *Human Ecology*, **32**(6), 735–73.

Krausmann, Fridolin and Helmut Haberl (2002), 'The Process of Industrialization from the Perspective of Energetic Metabolism. Socioeconomic Energy Flows in Austria 1830–1995', *Ecological Economics*, **41**(2), 177–201.

Krausmann, Fridolin, Heinz Schandl and Niels B. Schulz (2003), *Vergleichende Untersuchung zur langfristigen Entwicklung von gesellschaftlichem Stoffwechsel und Landnutzung in Österreich und dem Vereinigten Königreich*, Stuttgart: Breuninger Stiftung.

Krausmann, Fridolin, Helmut Haberl, Niels B. Schulz, Karl-Heinz Erb, Ekkehard Darge and Veronika Gaube (2003), 'Land-use Change and Socio-economic

Metabolism in Austria. Part I: Driving Forces of Land-use Change: 1950–1995', *Land Use Policy*, **20**(1), 1–20.

Maddison, Angus (2001), *The World Economy. A Millennial Perspective*, Paris: OECD.

Müller, Karl H. and Pavle Sicherl (2004), 'Time–distance Analysis: Method and Applications', *Wisdom*, (2a), 11–31.

Netting, Robert M. (1993), *Smallholders, Householders. Farm Families and the Ecology of Intensive, Sustainable Agriculture*, Stanford, CA: Stanford University Press.

Odum, Eugene P. (1959), *Fundamentals of Ecology*, Philadelphia: Saunders.

Oesterdieckhoff, Georg W. (2000), *Familie, Wirtschaft und Gesellschaft in Europa. Die historische Entwicklung von Familie und Ehe im Kulturvergleich*, Stuttgart: Breuninger Stiftung (Der europäische Sonderweg; 6).

Offe, Claus (1984), *'Arbeitsgesellschaft' – Strukturprobleme und Zukunftsperspektiven*, Frankfurt am Main: Campus.

Pfister, Christian (2003), 'Energiepreis und Umweltbelastung. Zum Stand der Diskussion über das "1950er Syndrom"', in Wolfram Siemann (ed.), *Umweltgeschichte Themen und Perspektiven*, München: C.H. Beck, pp. 61–86.

Pomeranz, Kenneth (2000), *The Great Divergence: China, Europe, and the Making of the Modern World Economy*, Princeton, NJ: Princeton University Press.

Schandl, Heinz and Nina Eisenmenger (2006), 'Regional Patterns in Global Resource Extraction', *Journal of Industrial Ecology*, **10**(4).

Schor, Juliet B. (2005), 'Prices and Quantities: Unsustainable Consumption and the Global Economy', *Ecological Economics*, **55**(3), 309–20.

Sieferle, Rolf P. (1995), 'Naturlandschaft, Kulturlandschaft, Industrielandschaft', *Comparativ* (4/1995), 40–65.

Sieferle, Rolf P. (1997), *Rückblick auf die Natur. Eine Geschichte des Menschen und seiner Umwelt*, München: Luchterhand.

Sieferle, Rolf P. (2001), *The Subterranean Forest. Energy Systems and the Industrial Revolution*, Cambridge: The White Horse Press.

Sieferle, Rolf P., Kurt Möser, Marcus Popplow, Nancy Kim and Otfried Weintritt (2004), *Transportgeschichte im internationalen Vergleich: Europa – China – Naher Osten*, Stuttgart: Der Europäische Sonderweg, Breuninger Stiftung, Band 12.

Siefere, Rolf P., Fridolin Krausmann, Heinz Schandl and Verena Winiwarter (eds) (2006), *Das Ende der Fläche. Zum Sozialen Metabolismus der Industrialisierung*, Wien: Böhlau.

Smil, Vaclav (1991), *General Energetics. Energy in the Biosphere and Civilization*, Manitoba, New York: John Wiley & Sons.

Smil, Vaclav (1994), *Energy in World History*, Boulder, San Francisco, Oxford: Westview Press.

Stiglitz, Josef E. (2002), *Globalization and its Discontents*, New York: W.W. Norton & Co.

UN (2004), *Energy Statistics Yearbook 2001*, New York: United Nations, Department of Economic and Social Affairs.

UN Statistics Division (2004), *National Accounts Main Aggregates Database*, http://unstats.un.org/unsd/snaama/Introduction.asp.

UNDP (2004), *Human Development Report 2004, Cultural Liberty in Today's Diverse World*, New York, NY: United Nations Development Programme.

Wallerstein, Immanuel (2000), 'Globalization or the Age of Transition. A Long-term View of the Trajectory of the World System', *International Sociology*, **15**(2), 249–65.

Webster, Frank (1995), *Theories of the Information Society*, London: Routledge.

Weisz, Helga, Marina Fischer-Kowalski, Clemens M. Grünbühel, Helmut Haberl, Fridolin Krausmann and Verena Winiwarter (2001), 'Global Environmental Change and Historical Transitions', *Innovation – The European Journal of Social Sciences*, **14**(2), 117–42.

Weisz, Helga, Fridolin Krausmann, Christof Amann, Nina Eisenmenger, Karl-Heinz Erb, Klaus Hubacek and Marina Fischer-Kowalski (2006), 'The Physical Economy of the European Union: Cross-country Comparison and Determinants of Material Consumption', *Ecological Economics*, **58**(4), 676–98.

Index